Organic Pollutants: Analysis and Remediation

Organic Pollutants: Analysis and Remediation

Kaleb Lynch

MURPHY & MOORE
www.murphy-moorepublishing.com

Organic Pollutants: Analysis and Remediation
Kaleb Lynch
ISBN: 978-1-63987-418-7 (Hardback)

MURPHY & MOORE
Published by Murphy & Moore Publishing,
1 Rockefeller Plaza,
New York City, NY 10020, USA

Cataloging-in-Publication Data

Organic pollutants : analysis and remediation / Kaleb Lynch.
p. cm.
Includes bibliographical references and index.
ISBN 978-1-63987-418-7
1. Organic compounds--Toxicology. 2. Organic compounds--Environmental aspects.
3. Environmental toxicology. 4. Pollutants. 5. Organic wastes. I. Lynch, Kaleb.
TD196.O73 O74 2022
577.278--dc23

For more information regarding Murphy & Moore Publishing and its products, please visit the publisher's website www.murphy-moorepublishing.com

TABLE OF CONTENTS

PREFACE

This book is a culmination of my many years of practice in this field. I attribute the success of this book to my support group. I would like to thank my parents who have showered me with unconditional love and support and my peers and professors for their constant guidance.

Organic pollutants are the organic compounds that show resistance to environmental degradation through biological, chemical and photolytic processes. They may arise naturally like volcanoes and various biosynthetic pathways or can be man-made. Some examples of organic pollutants are dioxins, heptachlor, toxaphene, endrin, polychlorinated biphenyls and lindane. These are used as solvents, pharmaceuticals, pesticides and industrial chemicals. They have adverse impacts on human health like endocrine disruption, gestational weight gain, additive and synergistic effects and effects on reproductive systems. They may also affect the environment through two processes. These are long-range transport and bioaccumulation. This book is a valuable compilation of topics, ranging from the basic to the most complex theories and principles related to this field. Also included in this book is a detailed explanation of the various concepts and applications related to the field of organic pollutants. Through this book, we attempt to further enlighten the readers about the new concepts in this field.

The details of chapters are provided below for a progressive learning:

Chapter - Introduction

The organic compounds that are resistant to environmental degradation through chemical, biological, and photolytic processes are termed as organic pollutants. Pesticides, solvents and pharmaceuticals, and industrial chemicals are some of the organic pollutants used. This chapter has been carefully written to provide an easy understanding of organic pollutants and its related aspects.

Chapter - Types of Organic Pollutants

There are many types of organic pollutants which include aldrin, chlordane, dieldrin, endrin, polychlorinated biphenyl, dioxins, pentachlorobenzene, endosulfan, organochloride, etc. The topics elaborated in this chapter will help in gaining a better perspective of the different types of organic pollutants.

Chapter - Bioaccumulation and Biomagnification

The accumulation of pesticides and other chemicals in an organism is termed as bioaccumulation. Biomagnification refers to the high levels of concentration of toxins and chemicals in the tissue of the tolerant organism. This chapter closely examines bioaccumulation and biomagnification to provide an extensive understanding of the subject.

Chapter - Analysis and Measurement

The analysis and measurement of organic pollutants include the analysis of remote zones air quality, total organic carbon, chemical oxygen demand, biochemical demand, etc. All the methods related to the analysis and measurement of organic pollutants have been carefully analyzed in this chapter.

Chapter - Effects of Organic Pollutants

Organic pollutants have adverse health effects on humans which include cancer, immune system suppression, decrements in cognitive and neurobehavioral function, disruption of sex steroid and thyroid function, and various chronic diseases, such as hypertension, cardiovascular disease, and diabetes. This chapter discusses the effects of organic pollutants in detail.

Chapter - Remediation and Prevention

The restoration of contaminated environmental sites in such a way that it is no longer a threat to human health is called remediation. There are various bioremediation and prevention technologies used such as bioventing, land farming, bioreactor, composting, bioaugmentation and biostimulation. This chapter has been carefully written to provide an easy understanding of the subject of remediation and prevention of organic pollutants.

Kaleb Lynch

Introduction

CHAPTER 1

The organic compounds that are resistant to environmental degradation through chemical, biological, and photolytic processes are termed as organic pollutants. Pesticides, solvents and pharmaceuticals, and industrial chemicals are some of the organic pollutants used. This chapter has been carefully written to provide an easy understanding of organic pollutants and its related aspects.

ORGANIC POLLUTANTS

Organic pollutants include many insecticides and herbicides that have been used in agriculture and pest control. Dichlorodiphenyltrichloroethane (DDT) is a pesticide, highly effective in controlling mosquitos, which was banned in the USA in 1972, but is persistent and its metabolite (DDE) is found in the blood of many populations. Other pollutants were manufactured for use in various industries [e.g. polychlorinated biphenyls (PCBs), phthalates], and others, such as dioxin, are unintended by-products of manufacturing. Phthalates are plasticizers used in bottles, toys and personal care products. PCBs are a large group of similarly structured compounds with variation in toxicity and persistence in the environment and in the body. Some forms are very similar to dioxin. Polybromated diphenyl esters (PBDEs) are fire retardants added to a large variety of consumer items that leach into surrounding materials and now can be detected in many populations. Organic pollutants such as PCBs, dioxin and DDT are lipophilic; they are stored in fat cells and can be retained for years. They cross the placenta, and lactation is a significant source of exposure. They are also found in dietary items such as fish, meat and dairy products.

PCBs may affect endocrine function, physical growth, maturation and/or cognitive or behavioral development of children and youth. Evidence of the effects of PCBs in humans comes from two types of study: studies of acute poisoning, either food poisoning or an occupational accident, and studies of chronic low-level exposures, usually from ingestion of foods with slight but measurable contamination.

Ingestion of rice oil contaminated with a mixture of PCBs, dioxin and dibenzofurans poisoned thousands of adults and children in Japan in 1968 and in Taiwan in 1978–79, producing diseases called Yusho and Yucheng, respectively. Yusho/Yucheng infants have had higher rates of mortality and lower body weights at birth. Even children born long after their mothers were exposed to the contaminated oil were more often born prematurely and small at birth, probably because during gestation they were exposed to the toxicant mixture that had been stored in their mothers' fat tissue. Reduced postnatal growth also characterizes Yusho/Yucheng children.

Studies of children born to women exposed to smaller amounts of PCBs over a long period have also found growth deficits. A common but not universal finding is that birth weight is reduced in response to PCB exposure. The effect seems not to be due to prematurity but to less growth during gestation. Head circumference may also be reduced. Three studies have related reduced birth weight to fetal exposure to dioxin and dioxin-like compounds. PCB levels and BMI are related early in life and at puberty. DDE, a metabolite of the insecticide DDT, has been related to reduced birth weight and height in several studies, but with increased height and weight for height in others.

Not all studies agree. It is important when interpreting conflicting results to account for differences in exposure among the studies. Certainly one would expect a smaller effect or none at all when the exposure is very low, and it is difficult to ascertain how exposure levels compare across studies because measurement techniques have changed substantially and only the most recent studies measure the contaminant compounds in similar ways. In addition, different effects may stem from differences in the timing of exposure: prenatal versus postnatal. Children exposed to PCBs from maternal consumption of fish from the Great Lakes were significantly lighter at 4 years of age, though not at 11 years of age. The reduction at age 4 was related to the PCB level at birth, reflecting prenatal exposure, but not to their current PCB level. Another longitudinal study found greater height in girls at 5 years of age in a cohort in which there had been differences in size at birth.

Clearly, not all studies of humans agree as to the size and direction of effects that these toxicants have. This creates questions about the influence of differences in levels of exposure, differences in the timing of exposure (prenatal or postnatal) and differences in the mixtures of compounds to which the sample was exposed. Excepting occupational exposures, mundane exposures tend to be a mixture of several compounds. While we tend to group all these persistent organic pollutants as "toxic", we have learned not to expect them to have similar effects on growth.

We do know that the alterations in growth are best explained as due to interference with hormonal signaling. Further, we know that different compounds affect signaling differently depending on the timing of exposure and their structure as some are agonistic and some antagonistic. Many studies of non-human animals have shown that hormone activity can be altered following exposure to PCBs and related compounds. Thyroid hormone signaling is especially important for normal physical and mental growth and development. Neurological effects are the most consistently reported effects of chronic PCB exposure. A study, found the combination of reduced thyroxine levels and increased thyroid-stimulating hormone levels (this usually signals low thyroid activity) in relation to levels of persistent PCBs that reflect past exposure, but not in relation to PCBs more reflective of current exposure. This suggests that prenatal or neonatal exposure may be influential. In this same group, there were alterations in performance on some cognitive tests and particular effects on memory.

Studies of the effects of phthalates in children have intensified recently as the material is found in more products and more populations. Effects of phthalates on the thyroid have been reported. Growth and sexual development can also be affected.

Many types of organic pollutant structurally resemble sex steroids as well as thyroid hormones. The possibility of deranged hormonal signaling of sexual development and reproduction has been a controversial area of research, but there is now sufficient evidence that it occurs in humans. Several studies conducted by different research groups and with different populations have reported that exposure to some pollutants accelerates the development of puberty, although the finding is not universal.

Whether the effect is one induced by prenatal or postnatal exposure, or both, is not clear. Newborns were found to have significantly lower testosterone and estradiol levels in relation to their mothers' levels of dioxin, dibenzofuran (similar to dioxin) and dioxin-like PCBs. Follow-up of boys in the Yucheng cohort who had substantial exposure to dioxin-like compounds showed reduced testosterone levels and increased FSH levels but no difference in Tanner stage. Preliminary results from the same cohort found significantly reduced penile lengths, a possible effect of their exposure prenatally when critical sexual differentiation and development is occurring.

A related compound, PBDE, is a new pollutant of concern. In adult men it has been strongly and inversely associated with a measure of androgen, with LH and FSH, and positively with inhibin B and sex steroid binding globulin. This result and other similar findings strongly suggest that chemicals in our environment can affect levels of hormones directly involved in reproduction and development. In true experimental studies of non-human animals endocrine disruption is clearly evident and this establishes the biological plausibility of the associations seen in observational studies of human populations. Research on children and PBDE is just beginning. If the effects seen in adults are present in children, they certainly could affect growth and sexual maturation.

There is sufficient evidence to be concerned about PCB exposure in children and the fetus. Especially convincing are the controlled laboratory studies of higher primates and rodents that show reductions in growth and alterations in sexual development that are dependent on normally functioning hormonal systems. Although it is not known whether a low-level exposure will produce effects on child development, we do see that high exposures, such as those from food heavily contaminated with PCBs, can produce predictable effects, and the possibility that the fetus is especially sensitive to PCBs, even at low levels, remains a very viable hypothesis.

Organic Pollutants in the Marine Environment

It has long been known that specific toxins accumulate in the natural environment and in living organisms, causing damage to health. As a result, the use of many of these chemicals is now prohibited. However, new toxic substances that were not initially recognized as a threat are frequently detected in the environment. Polyfluorinated compounds (PFCs) are one current example. There is still no solution to this problem.

The Downside of Consumption

Chemical-based products are found in plastic casing for computers, in athletic-hall flooring, in rubber soles for jogging shoes – the applications are endless. In consequence, a very wide range of chemicals is used in industry today. According to the Organisation for Economic Co-operation and Development (OECD), approximately 100,000 different chemical substances are currently on the market worldwide. In Europe alone, approximately 10,000 chemicals are produced and marketed annually in amounts of greater than 10 tonnes. It is estimated that between 1 and 3 per cent of these chemicals are problematical. These environmentally relevant pollutants include heavy metals such as lead and mercury, which are released into the environment by the burning of oil, mining activities, and industrial emissions and effluents. Persistent organic pollutants, known as POPs, are another problematical substance category.

During the phocine distemper virus (PDV) epidemics in 1988 and 2002, thousands of dead seals were washed up on German beaches and had to be collected and destroyed.

Poisonous and Persistent – POPs

The Stockholm Convention on Persistent Organic Pollutants adopted in 2001 deals with persistent organic pollutants (POPs), i.e. substances that possess toxic properties and resist degradation. They include pesticides such as DDT and lindane, industrial chemicals such as polychlorinated biphenyls (PCBs), and substances such as dioxins, which are the unwanted by-products of manufacturing and combustion processes. As these substances are highly stable and therefore non-degradable to a large extent, they can be transported over long distances and accumulate in the environment.

POPs cause problems because they are stored in the fatty tissue or organs of animals, where they can have toxic effects. For example, they can disrupt the endocrine system, cause cancer or genetic defects, and weaken the immune system.

Various effects of POPs on marine mammals have been investigated. Studies of Baltic ringed seals (Phoca hispida) and grey seals (Halichoerus grypus) found uterine occlusions, stenoses and tumours, resulting in reduced reproductive ability. Other observed effects included colonic ulcers, as well as reduced bone density, which led to changes in the skeletal system. In seals and porpoises, researchers found indications that POPs depress the immune and endocrine systems. A further topic of discussion in this context is whether these pollutants and the weakening of the immune system affect the spread of epidemics, such as the disease that killed thousands of seals in the North Sea in 1988 and again in 2002 – probably an epidemic of the phocine distemper virus. Humans mainly ingest POPs from food and drinking water, but also from the air (mainly by breathing in dust particles) and through the skin (through direct contact with the chemicals). The highest concentrations of POPs are generally found in marine mammals and humans, both of which are at the top of the food chain.

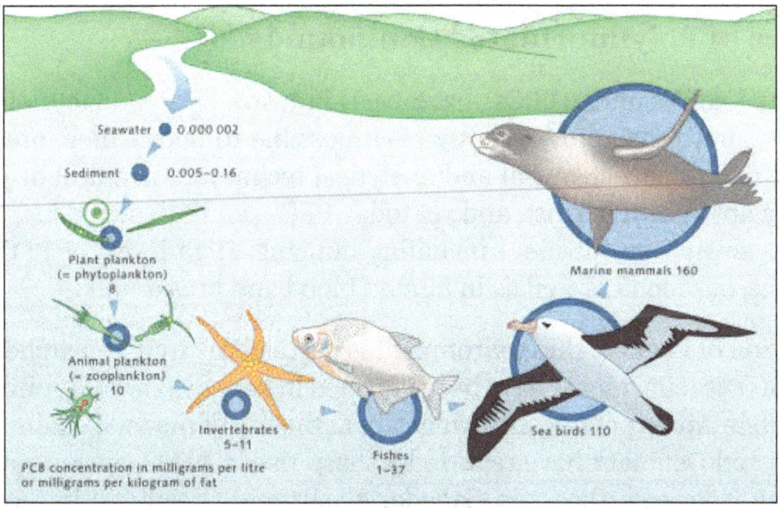

Bioaccumulation of toxins in the marine food chain has long been recognized as a problem. The process illustrated here relates to polychlorinated biphenyls (PCBs), a typical environ-mental toxin.

Polyfluorinated Compounds: A Fresh Cause for Concern

New types of persistent toxic compounds of non-natural origin were identified in the environment at the end of the 1990s, which could not be detected before as the appropriate technology and analytical methods had not yet been developed. These include polyfluorinated compounds (PFCs), which have been used in a wide variety of every-day applications for more than 50 years. PFCs are mainly used as fluoropolymers in the textile industry, for example, in the manufacture of breathable membranes for outdoor clothing, and in the paper industry in the production of water-, stain- and grease-proof paper (e.g. fast-food packaging). They are also used for surface treatment of furniture, carpets and clothing textiles and in non-stick coatings for cookware (such as Teflon frying pans).

It is believed that a total of six manufacturers have produced around 4500 tonnes of PFCs every year over the past decade: a relatively small amount compared with other chemicals. This group of substances is significant nonetheless, due to its environmentally relevant properties, as some PFCs are highly bio-accumulative in organisms.

At present, more than 350 different PFCs are known to exist. The best-known is perfluorooctanesulfonic acid, more commonly known as PFOS. Based on animal experiments with PFOS, researchers conclude that repeated exposure can have an extremely adverse effect on human health; among other possible effects, it may cause damage to the liver. PFOS may also be carcinogenic, and it is also thought to impair the development of progeny. PFOS therefore recently became the first PFC to be listed as a persistent organic compound (POP) under the Stockholm Convention, which means that it is now on the list of particularly hazardous chemicals for which a worldwide ban is to be imposed.

Occurrence of Polyfluorinated Compounds

Polyfluorinated compounds (PFCs) have been industrially manufactured for around half a century, but it has only recently been possible to detect their presence in the environment due to new chemical and analytical techniques. Natural origins of these chemicals are not known to exist, and yet today PFCs can be detected in water, soil, air and living organisms worldwide – including humans. High levels of PFCs have been found in numerous foods as well as in human blood and breast milk.

The distribution of PFOS in the environment is particularly well-researched. High concentrations of these substances have been detected in fish, seals and sea birds worldwide and, above all, in Arctic polar bears, which are at the top of the food chain. Researchers from Canada and Denmark have reported a sharp rise in PFOS concentrations in liver samples taken from polar bears in Canada, Alaska and Greenland in recent decades. Compared to other environmentally relevant POPs, such as polychlorinated biphenyls, PFCs are found in much higher concentrations. In Swedish studies of human blood from 1994 to 2000, the mean PFC concentration was 20 to 50 times higher than the concentration of the polychlorinated biphenyls and about 300 to 450 times higher than that of hexachlorobenzene, two "classic" organic pollutants that have been recognised as hazardous for decades.

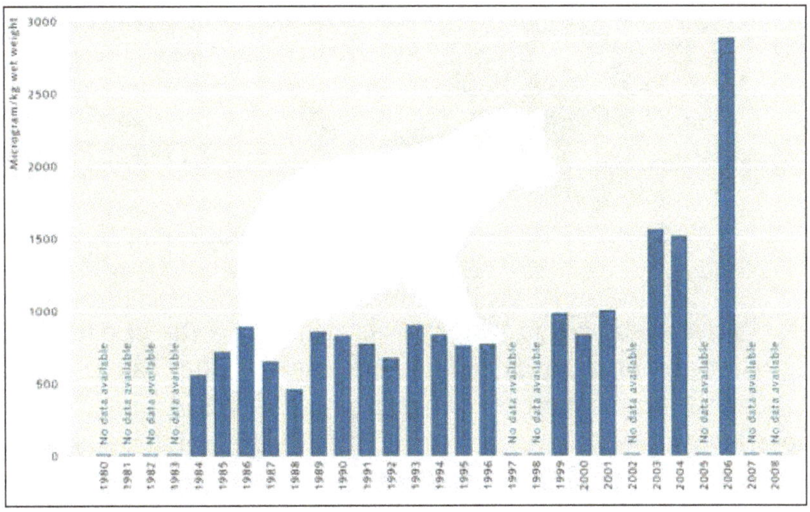

PFOS concentrations in the livers of East Greenland polar bears have increased significantly in recent years. The measurements were obtained from deep-frozen liver samples.

Transport of PFCs

The detection of PFCs and especially PFOS in marine mammals such as Arctic seals and polar bears, and even in the blood of the Arctic's human inhabitants, the Inuit, raises the question of transportation: How did these substances end up in the sea and even in the Arctic? There are numerous different sources of PFCs. They are released, for example, during the use of the everyday consumer durables mentioned above – from carpeting, outdoor clothing, cookware and fast-food packaging. However, in Germany, relatively large concentrations of PFCs also enter the rivers from municipal and industrial wastewater treatment plants, which cannot capture these substances. The rivers then wash these substances into the North Sea. From here, they are carried by the main North Sea and Atlantic Ocean currents to the Arctic, where they are ingested by microorganisms in the water and thus enter the food chain, bioaccumulating in larger organisms and finally in the organs of polar bears and humans.

PFCs are also transported long distances through the atmosphere by the movement of air masses. Compounds such as PFOS are not volatile, but volatile precursor compounds escape into the atmosphere during the manufacturing process. Physical and chemical processes that take place in the atmosphere then convert these chemical precursors into stable end products such as PFOS. These are removed from the air by precipitation and enter the seawater in soluble form or bound to dust particles, or are deposited on land or ice surfaces. PFCs can thus travel great distances and can be detected in the environment a long way from their place of origin or use.

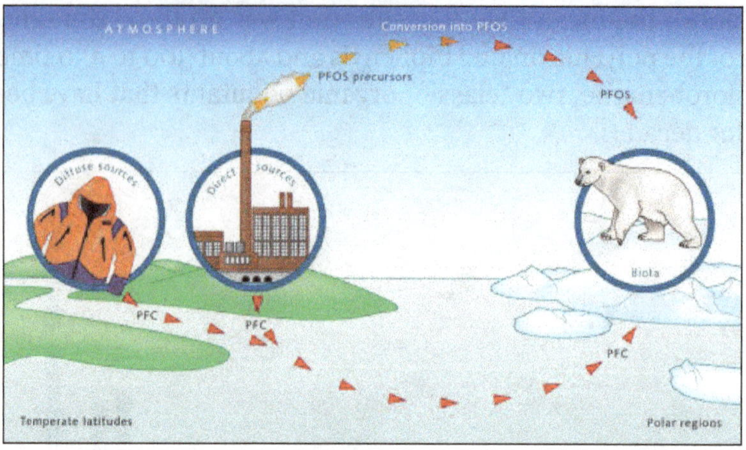

PFCs can travel great distances in water or air. Through a direct pathway, they enter the rivers in wastewater and are carried down to the sea. They can also be transported indirectly through the atmosphere. For example, volatile PFOS precursors are released into the atmosphere, where they are converted into PFOS, which is then deposited back on the Earth's surface at the place of origin or elsewhere in rainfall or in dust particles.

Protection from New Pollutants

Today PFCs are distributed all over the world. They are found in water, in the air, in living organisms and even in our own bodies. They are likely to persist for generations. This group of substances clearly shows that it is impossible to predict all the environmental impacts, or the delayed effects, of new chemical substances. In the future too, it is likely that some substances that were initially regarded as harmless, but whose undesirable effects can only be discerned after some time has elapsed, will be detected in the marine environment. Nowadays, however, intensive efforts are being made to limit the further global spread of pollutants. For example, risk assessments are carried out before chemicals are licensed for use, in order to determine to what extent they could constitute a hazard. There are also various voluntary renunciation schemes for producers, as well as relevant legislation. In other words, a start has been made.

PERSISTENT ORGANIC POLLUTANTS

Persistent Organic Pollutants (POPs) are toxic substances composed of organic (carbon-based) chemical compounds and mixtures. They include industrial chemicals like PCBs and pesticides like DDT.

They are primarily products and by-products from industrial processes, chemical manufacturing and resulting wastes. The existence of POPs is relatively recent, dating to the boom in industrial production after World War II.

Why are POPs a Threat?

Today, POPs are found almost everywhere - in our food, soil, air and water. Wildlife and humans around the world carry amounts of POPs in their bodies that are at or near levels that can cause injury.

POPs pose a particular hazard because of four characteristics: they are toxic; they are persistent, resisting normal processes that break down contaminants; they accumulate in the body fat of people, marine mammals, and other animals and are passed from mother to fetus; and they can travel great distances on wind and water currents. Even small quantities of POPs can wreak havoc in human and animal tissue, causing nervous system damage, diseases of the immune system, reproductive and developmental disorders, and cancers.

Of the numerous POPs that are prevalent in our environment, 12 of the most persistent, bioaccumulative chemicals have been identified for priority action. These 12 POPs are targeted in a new international treaty. The Stockholm Convention on Persistent Organic Pollutants will phase out and eliminate the production and use of those chemicals, as well as new ones that would be added once the treaty is in force.

The 12 targeted POPs include eight pesticides (aldrin, chlordane, DDT, dieldrin, endrin, heptachlor, mirex, and toxaphene), two types of industrial chemicals (polychlorinated biphenyls or PCBs and hexachlorobenzene), and two chemical families of unintended by-products of the manufacture, use, and/or combustion of chlorine and chlorine-containing materials (dioxins and furans). All 12 targeted POPs are also endocrine disruptors' Chemicals that can interfere with the body's own hormones. Endocrine disrupting chemicals can be hazardous at extremely low doses and pose a particular danger to those exposed in the womb. During prenatal life, endocrine disruptors can alter development and undermine the ability to learn, to fight disease, and to reproduce.

Where are Persistent Organic Pollutants found?

POPs are found everywhere in the world in measurable amounts.

- Food: Fish, shellfish, or wild foods in which POPs have bioaccumulated.
- Air: Indoors and outdoors, cigarette and secondhand smoke, and vehicle exhaust.
- Consumer products: Pesticides, insecticides, cigarettes, and some paints.

Persistent Organic Pollutants (POPs) commonly enter(s) the body through:

- Ingestion (swallowing): Swallowing food or water contaminated with POPs.
- Inhalation (breathing): Breathing indoor or outdoor air contaminated with POPs, vehicle exhaust, or cigarette and secondhand smoke.
- Skin contact: Touching products made with POPs.

Exposure to certain POPs can cause various effects:

- Greater susceptibility to disease,
- Contaminated breast milk,
- Damage to the immune, neurological, and reproductive systems,
- Endocrine disruption,
- Decreased comprehension,
- Birth defects,
- Cancer,
- Death.

Who is at Risk for Exposure to Persistent Organic Pollutants?

- Subsistence fishers and hunters: Fish, shellfish, and wild foods that are locally obtained may have higher levels of POP contamination.
- Women and men of child-bearing age: Exposure to POPs can harm human reproductive systems and a developing fetus.
- Children and the elderly: Sensitive populations are more vulnerable to POPs.

Persistent organic pollutants (POPs) are organic chemical compounds that present toxic properties, resistant to degradation, bioaccumulate and are transported by air, water and migratory species across international boundaries and deposited far from their place of origin.

Being volatile substances, POPs evaporate into the air in warm regions of the globe, are transported by air currents up to cold regions and in mountainous regions where they condense. Thus, the Polar Regions are becoming areas where the concentration of POPs is expected to be high due to the global process of condensation. Due to the remote transport phenomena, pollution by these substances affects large-scale systems.

Over time, POPs are at risk of inducing adverse effects on human health and the environment.

POPs include the following classes of substances: organochlorinated pesticides (OCP), polychlorinated biphenyls (PCBs), perfluorated compounds (PFCs), brominated compounds (BFR), dioxins and furans. Most of these substances are of anthropogenic origin, but substances such as dioxins and furans may have natural origin such as volcanism or vegetation fires vegetaţie. POPs are mostly man-made chemical products intended to be used in agriculture (pesticides) and in industry (polychlorinated biphenyls, hexachlorobenzene, brominated compounds, perfluorated compounds) or unintentional by-products resulting from industrial processes, or from waste incineration (dioxins and furans).

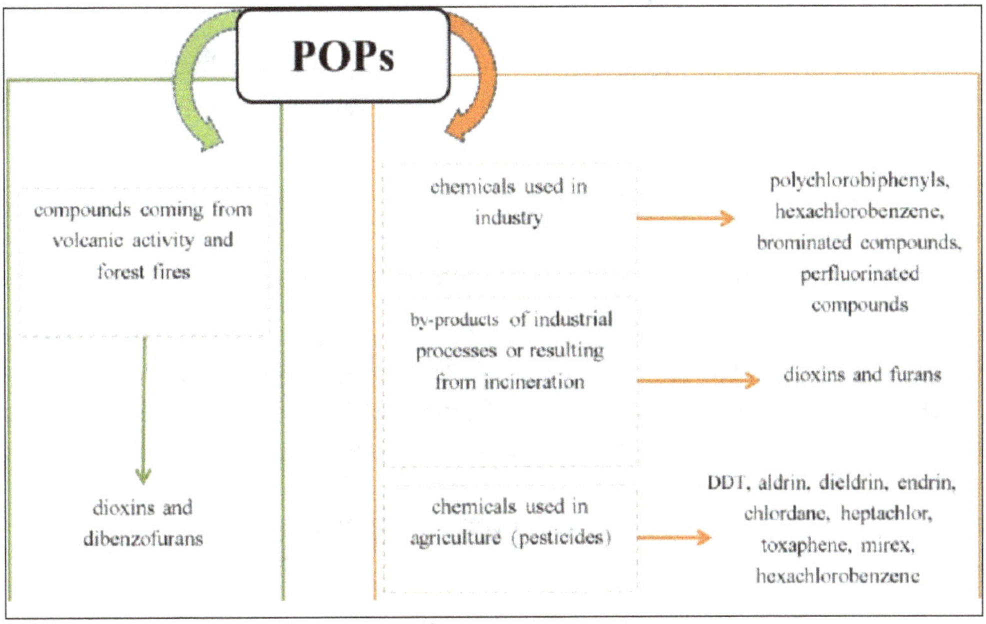

Classification of persistent organic compounds according to their origin.

The main anthropogenic POP pollution sources are:

- Intensive agriculture and zootechnics.
- City dumps and landfills, uncontrolled degradation of domestic waste.
- Waste from the destruction and elimination of pesticides producing facilities, electrical power plants and their equipment.
- Electronic waste recycling facilities.
- Waste incineration.
- Pesticide industry (including waste generated by this industry).
- Chemical and petrochemical industry.
- Roads and railroads – due to motor vehicles and trains.
- Water and waste water treatment plants.
- Electrical power plants.

Most POP chemicals are non-polar organic compounds, consequently hydrophobic, with extremely low water solubility; they are lipophilic, which causes them to accumulate in the organisms' fat tissues. Due to the bioaccumulation and biomagnification phenomena, the POP concentration may be much higher in the tissues of the organisms found at the high end of the trophic chains (fish and mammal species, birds and humans) than in the environment.

In the case of aquatic ecosystems, the chemicals are adsorbed on the sediment particles or on the solid particles suspended in the water column, especially on organic particles and they accumulate in the aquatic organisms' fat tissues.

The extensive contamination with POPs of various terrestrial or aquatic environments and the bioaccumulation and biomagnification processes determine chronic and acute toxic effects on organisms, including on the human population.

In 1995, the United Nations Environmental Programme (UNEP) called to global scale action in the field of POPs. The International Forum for Chemical Security (IFCS) and the International Programme for Chemical Security (IPCS) performed a risk assessment for human health and for environment for 12 chemicals considered the most dangerous, the so-called „dirty dozen": aldrin, chlordane, DDT, dieldrin, endrin, heptachlor, mirex, toxaphene, hexachlorobenzene, polychlorobiphenyls, polychloro-dibenzo-pdioxins, polychloro-dibenzo-furans.

In 2001, the Stockholm Convention on persistent organic pollutants was signed, with the purpose of protecting human health and environment against the negative effects of POPs by guaranteeing the safe phasing out of these chemicals as well as the reduction of their production and use. The Stockholm Convention entered into force; it is signed by 152 states and ratified by 178 states.

The Convention's Annexes list the chemicals the Convention applies to; Annex A refers to the substances for which are required phasing out measures for production and use, Annex B regards the chemicals for which are required active limitations of production and use, and Annex C includes the chemicals for which are required measures for reducing the accidental release into the environment.

At the time of its adoption, the Convention applied to the 12 chemicals of the "dirty dozen". In 2009, 2011, 2013, and 2015 the Conference of the Parties adopted amendments to Annexes A, B, and C of the Convention, by adding new POPs.

In order to assess the accomplishment of the Conventions' objectives, the Conference of the Parties launched the Global POP Monitoring Plan, a plan that creates a unified framework for the collection of comparable monitoring data on the POPs presence in all regions, with the purpose of identifying the dynamics of these substances concentrations in time and, also, to assess the POPs transport in the environment at regional and global scale. The monitoring activities are focused on measuring the POPs in the air, the human breast milk and human blood the water-soluble POPs (perfluorooctane sulfonic acid – PFOS and its salts, perfluorooctane sulfonyl fluoride - PFOSF) are monitored in the surface waters.

According to the European Commission report on the POPs, the European countries with high emissions of PAH (polycyclic aromatic hydrocarbons), PCDD/PCDF (polychlorodibenzo-p dioxins/ polychlorinated dibenzofurans), and HCB (hexachlorobenzene), were Spain, Italy, Portugal, Poland and United Kingdom. From the listed emissions, the maximum values were recorded for Poland, for PAH with 166 tons/year, Portugal for PCDD/PCDF with 574 grams/year and Spain with 7678.7 kg/year for HCB. The minimum values were recorded for Luxembourg both for PAH and PCDD/PCDF with 1.4 tons/year and 1.6 grams/year respectively, and Ireland for HCB with 0.1 kg/year.

The data presented by the countries that are Parties to the Stockholm Convention regarding the trend in the POP emissions since 1990 to 2006 show that PAH emissions increased for Italy, Latvia and Denmark, but the other countries showed a stable evolution or a decrease in these compounds emissions. In the case of PCDD/PCDF emissions, growing values appear for Latvia, Lithuania and Romania, in Romania's case the growth being more obvious starting with 2000. Fortunately, in the case of these pollutants most of the countries showed decreasing trends, a very important thing since PCDD/PCDF is classified between the most toxic POP and group 1 carcinogenic for humans. In the case of HCB emissions, one can notice an increase for Spain, Belgium, Luxembourg and Latvia, most of the other states showing a decreasing trend of emissions.

International studies showed that the continental aquatic ecosystems are vulnerable to POP pollutions and the accumulation of these substances in fish meat that can be consumed by humans is significant. Thus being the case, it is necessary to have a mechanism for coherent data collection at regional and global scale on the concentration of these chemicals in water, sediment and aquatic organisms; also research efforts are required in order to evaluate the POP effects on ecological processes, the evaluation of toxicological and non-toxic effects, and of the medium and long term risks.

POPs have long half-lives in soils, sediments, air and biota. No consensus exists on how long the half-life in a given medium should be for the word "persistent" to be applicable; nevertheless, the half-life of a POP might be years or decades in soil/sediment and a number of days in the atmosphere. Several thousand POP chemicals are known. Many of them are members of a definite series or "families" of chemicals. For instance, 209 diverse groups of polychlorinated biphenyls have been identified; they differ from each other by the level of chlorination and the substitution position. The carbon-chlorine bond is very stable toward hydrolysis and larger numbers of chlorine substitution and/or functional groups lead to greater resistance to biological and photolytic degradation. Because POPs break down very slowly, they will be present in the environment for a long time, even if all new sources are immediately eliminated.

POPs in the environment are transported at low concentrations by movement of fresh and marine waters; in addition, because they are semi-volatile, POPs are transported over long distances in the atmosphere. The result is widespread distribution of POPs across the globe, including regions where they have never been used. The POP levels observed in the Arctic region are surprising to many people because some of these pollutants have been banned from the USA and Canada for many years. POPs travel toward colder areas such as Alaska and then sink because of the colder temperatures. The settled contaminants remain in the area for a long period because the temperature does not allow them to easily breakdown. Consequently, they move from the air and water to soil and plants and then to animals and humans with ease. The persistence of contaminants in the Arctic from distant sources first came to light in the late 1970s when pesticides were found in polar bear fat tissue. The reality of atmospheric POPs and their effects on wildlife and human health then became evident. In addition to the Arctic, researchers have begun to search for evidence of airborne POPs in other cold ecosystems and mountain environments.

POPs are usually hydrophobic (i.e., "water-hating") and lipophilic (i.e., "fat-loving") chemicals. In marine and terrestrial systems, they bind strongly to solids, particularly organic matter, evading the aqueous segment. They also enter the lipids of organisms more easily than the inside of the aqueous medium of cells and are stockpiled in fatty tissue. This stockpiling in fatty tissue allows the compounds to persevere in biota, where the metabolism rate is low. Consequently, POPs may climb the food chain. POPs tend to move in the gas phase under environmental temperatures. Therefore, they may volatilize from soils, vegetation, and aquatic systems into the air and, because of their resistance to breakdown reactions in air, migrate long distances before being re-deposited. The sequence of volatilization and condensation can be repeated frequently; as a consequence, POPs are detected in regions far removed from where they were used or discharged. POPs can partition between particles and aerosols in the air and are reliant on ambient temperature and the physico-chemical characteristics of the substances. The combination of permanency and the tendency to form a gas under suitable environmental conditions causes POPs to migrate long-range atmospheric distances. The combined effects of confrontation, metabolism, and lipophilicity result in the accumulation of POPs in food chains.

Sources and Emission

Fire is the foremost source of POPs, including natural, unintentional, and planned burning of flora. POPs are recalcitrant in all components of the environment. They pass into the air from various industrial sources (e.g., incinerating plants, power stations, and heating stations), household furnaces, volatilization from water surfaces and soil, landfill transport, and consumption of agricultural sprays. Additional sources of POP contaminants include unpremeditated generation resulting from chemical amenities, incineration, assorted combustion processes such as forest fires, and decomposition of wastes containing PCBs. This group of wastes can result in several zones and halting from diverse activities, such as the consumption of obsolescent oil, the fixing and repair of tools, the destruction of buildings, cement manufacturing, evaporation, the burning of animal remains, coal ignition, lixiviation of dumps and reprocessing actions, municipal incineration, therapeutic waste, organochlorine pesticide plants, industrial chlor-alkali plants, aluminum secondary plants, furnace and foundry coke production plants, sewage sludge, hazardous waste/plastic waste in landfills, organochlorine pesticide stowage, and fly ash stowage.

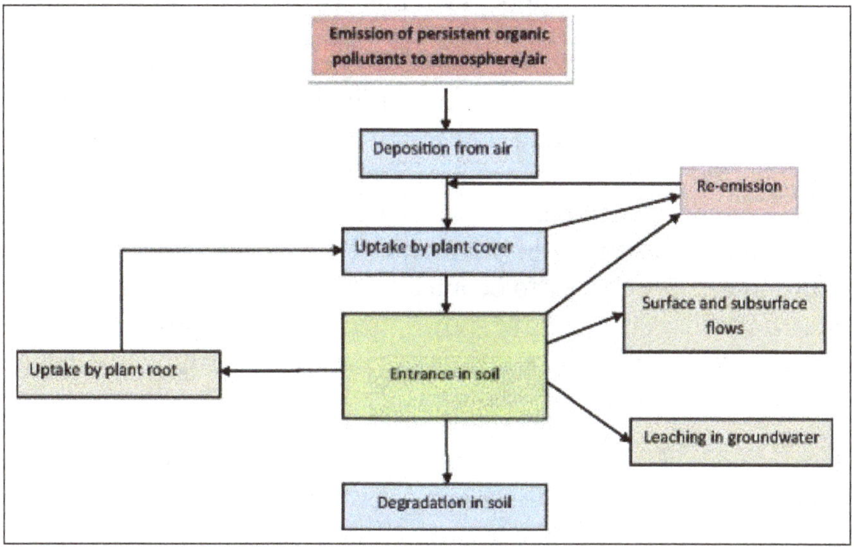

Conceptual model for the behavior of persistent organic pollutants in the air–plant–soil system.

Elements, liquid fuels, oils, soil, fats, sediment, and ash in the aquatic environment originate from wastewaters from manufacturing plants or POP consumption, in consort with overflows from roads and fields and from atmospheric deposition. Seas and oceans are prevalent reservoirs where POPs collect from river sediments, by atmospheric deposition, and unintentionally. They are deposited in sediments on the beds of oceans, seas, and large lakes, where they can be discharged after a time and then withdraw into the air.

Dosage and Toxicity of POPs

POPs cause concern because of their potential toxicity and predominance in environmental media, even at distant topographical localities. The adverse effects of POPs on the ecosystem have triggered considerable worry in recent years, which has resulted in the monitoring and sanctioning of the consumption of these toxic compounds in several states. A large variety of organic compounds contaminate water bodies, which is a foremost concern that has thus far been focused on PCBs and organochlorine pesticides (OCPs). The continuous increase of these organic contaminants in the aquatic environment is largely attributable to their high persistence and resistance to natural biodegradation.

Public concern about the toxicity by POPs has been amplified recently because a number of these chemicals are recognized as hormone disrupters, which can cause malfunctioning of the endocrine and reproductive systems in animals and humans. These chemicals impose serious health effects on our ecosystem, and none of them has any beneficial effects. After these contaminants have been introduced into the ecosystem, they continue to exist in the environment for several years, causing complications such as learning disabilities, birth defects, cancer, and behavioral, neurological, reproductive, and immunological disorders in wildlife species and humans.

Recently, certain POPs have also been implicated in immunodeficiency in infants and children and in the intensification of infection, in combination with growth deformities, neurobehavioral deficiency, and cancer and tumor generation or elevation. Several authors have reported that POPs are substantial risk factors in the etiology of human breast cancer. Because children are still developing, they are substantially more susceptible than adults to the effects of pollutants. Their developing cells are sensitive to contaminants and are more likely to be affected by exposure to POPs. The brain is the greatest concern because studies have shown that children exposed to POPs during infancy scored remarkably lower on assessments used to determine intelligence and the ability to shut out distractions.

POPs can have serious side effects, including birth defects, certain cancers and tumors at multiple sites, immune-system disorders, reproductive problems, reduced ability to fight diseases, stunted growth, and permanent impairment of brain function. POPs are suspected carcinogens and have been implicated in diseases such as endometriosis (a painful, chronic gynecological disorder that affects uterine tissues), increased incidence of diabetes, and neurobehavioral impairment, including learning disorders, reduced performance on standard tests, and changes in temperament. Numerous methods have been published over the past 30 years related to specific analytical techniques for the determination of PCBs and OCPs in food and environmental matrices.

STOCKHOLM CONVENTION ON PERSISTENT ORGANIC POLLUTANTS

Stockholm Convention on Persistent Organic Pollutants (POPs) is an international treaty to protect human health and the environment from the harmful effects of POPs. The Convention was adopted on 22 May 2001 in Stockholm, Sweden and entered into force.

More than 170 countries have ratified the Convention up to date. The Convention requires that Parties to the convention take measures to eliminate or restrict the production and use of certain hazardous chemicals on the List of Persistent Organic Pollutants (POPs) in the Convention.

Items	Content
Target	Persistent Organic Pollutants(POPs)
List Available?	Yes
Control Measures	Eliminate or restrict the production and use
Articles Affected?	Yes

List of Persistent Organic Pollutants (POPs) in the Convention

Currently there are over 30 chemicals on the List of Persistent Organic Pollutants (POPs) in the Convention. The List consists of three annexes:

- Annex A (Elimination): The production and use of chemicals on annex A must be eliminated unless there are specific exemptions.
- Annex B (Restriction): The production and use of chemicals on annex B must be restricted.
- Annex C (Unintentional Production): Measures must be taken to reduce the unintentional releases of chemicals on Annex C.

Annex A:

- Aldrin,
- Chlordane,
- Chlordecone,
- Dieldrin,
- Endrin,
- Mirex,
- Heptachlor,
- Hexabromobiphenyl,
- Hexabromocyclododecane (HBCD),
- Hexabromodiphenyl ether and heptabromodiphenyl ether,
- Hexachlorobenzene (HCB),
- Alpha hexachlorocyclohexane,
- Beta hexachlorocyclohexane,
- Lindane,
- Pentachlorobenzene,

- Polychlorinated biphenyls (PCB),
- Technical endosulfan and its related isomers,
- Tetra bromodiphenyl ether and pentabromodiphenyl ether,
- Toxaphene,
- Chlorinated naphthalenes,
- Hexachlorobutadiene,
- Pentachlorophenol,
- ecabromodiphenyl ether (commercial mixture, c-decaBDE),
- Short-chained chlorinated paraffins,
- Dicofol,
- Perfluorooctanoic acid (PFOA) its salts and PFOA-related compounds.

Annex B:

- DDT: can be used for disease vector control only.
- Perfluorooctane sulfonic acid, its salts and perfluorooctane sulfonyl fluoride: can be used for certain accepted purposes.

Annex C:

- Polychlorinated dibenzo-p-dioxins ("dioxins"),
- Polychlorinated dibenzofurans (PCDF) ("furans"),
- Hexachlorobenzene (HCB),
- Pentachlorobenzene,
- Polychlorinated biphenyls (PCB),
- Hexabromocyclododecane(HBCD).

It shall be noted that:

- 3 chemicals are listed in both Annex A and Annex C: Hexachlorobenzene (HCB), Pentachlorobenzene and Polychlorinated biphenyls (PCB).
- Even for chemicals listed on Annex A, there may be some specific exemptions. For example, Hexabromocyclododecane(HBCD) can be used in expanded polystyrene and extruded polystyrene in buildings.
- Many chemicals have multiple CAS numbers. For example, there are 2 CAS numbers for Hexabromocyclododecane(HBCD): CAS25637-99-4 and CAS3194-55-6.

Impacts of the Stockholm Convention

The Stockholm Convention has a big impact on both chemical industry and other sectors which use certain hazardous substances (for example, flame retardants) in their articles and parts. Once a hazardous substance is added to Annex A, it will face a global ban. Companies must take measures to eliminate or substitute the hazardous substance in their products.

Countries that have ratified the Convention usually have their own environmental regulations to eliminate or restrict the production and use of listed Persistent Organic Pollutants (POPs) in their own jurisdictions.

DEGRADATION PATHWAYS OF PERSISTENT ORGANIC POLLUTANTS IN THE ENVIRONMENT

There are many ways that release POPs to the environment. POPs such as pesticides are released, as a result of plant protection efforts. The main sources of pesticide pollution include their use, transportation, storage, and disposal of obsolete pesticides. Some other chemicals [polychlorinated biphenyls (PCBs), pentachlorophenol] are used as oils, as dielectric and cooling fluids in capacitors and transformers, for wood preservation, etc., and are released into the environment as a result of spills and evaporation. A number of substances like dioxins/furans, polycyclic aromatic hydrocarbons (PAHs), and hexachlorobenzene are by-products of many industrial processes, mainly, thermal (fuel combustion and waste incineration, ferrous industry, coke and aluminum production, road transport, chemical synthesis of chlorinated substances, etc.) and are emitted directly into the air.

Emission of dioxins from industrial wastes.

Why are POPs Resistant to most of the Environmental Degradation Processes?

In the initial list of the Stockholm Convention, POPs are typically polyhalogenated organic compounds which exhibit high lipid solubility. For this reason, they bioaccumulate in fatty tissues of animals. Halogenated compounds also exhibit great stability reflecting the nonreactivity of C-Cl bonds toward hydrolysis and photolytic degradation in the environment. The high electronegativity of halogens compared to carbon gives the carbon-halogen bond a significant polarity/dipole moment. The electron density is concentrated around the halogen, leaving the carbon relatively electron poor. This introduces ionic character to the bond through partial charges ($C^{\delta+}Cl^{\delta-}$). The partial charges on the chlorine and carbon exert electrostatic attraction between them, contributing to the unusual bond strength of the carbon-halogen bonds. The carbon-halogen bonds are known to be "the strongest in organic chemistry," because halogen forms the strongest single bond to carbon. The bond also strengthens and shortens as more halogens are added to the same carbon on a chemical compound.

The stability and lipophilicity of persistent organic compounds often correlate with their halogen content; thus, polyhalogenated organic compounds are of particular concern. They exert their negative effects on the environment through two processes, long-range transport, which allows them to travel far from their source, and bioaccumulation, which reconcentrates these chemical compounds to potentially dangerous levels in animal tissues.

Health Effects of POPs

POP exposure may cause developmental defects, chronic illnesses, and death. Some are carcinogens, possibly including breast cancer. Many POPs are capable of endocrine disruption within the reproductive system, the central nervous system, or the immune system. People and animals are exposed to POPs mostly through their diet, occupationally, or while growing in the womb. For humans not exposed to POPs through accidental or occupational means, over 90% of exposure comes from animal food products due to bioaccumulation in fat tissues and bioaccumulation through the food chain. In general, POP serum levels increase with age and tend to be higher in females than males.

Effect on Endocrine Disruption

Exogenous substance/mixture that alters the function(s) of the hormonal system and consequently causes adverse effects in an intact organism or its progeny or its subpopulation is called endocrine disruptors. The majority of POPs are known to disrupt the normal functioning of the endocrine system. Low-level exposure to POPs during critical developmental periods of the fetus, newborn, and child can have a lasting effect throughout its lifespan. A 2002 study synthesizes data on endocrine disruption and health complications from exposure to POPs during critical developmental stages in an organism's lifespan. The study aimed to answer the question whether or not chronic, low-level exposure to POPs can have a health impact on the endocrine system and development of organisms from different species. The study found that exposure of POPs during a critical developmental time frame can produce a permanent change in the organism's path of development. Exposure of POPs during non-critical developmental time frames may not lead to detectable diseases and health complications later in their life. In wildlife, the critical development time frames are in utero, in ovo, and during reproductive periods. In humans, the critical development time frame is during fetal development.

Effect on the Reproductive System

The same study in 2002 with evidence of a link from POPs to endocrine disruption also linked low-dose exposure of POPs to reproductive health effects. The study stated that POP exposure can lead to negative health effects especially in the male reproductive system, such as decreased sperm quality and quantity, altered sex ratio, and early puberty onset. For females exposed to POPs, altered reproductive tissues and pregnancy outcomes as well as endometriosis have been reported.

Effect on Gestational Weight Gain and Newborn Head Circumference

The lower the birth weight and head circumference of the infants was, the higher POPs levels during prenatal development had been, but only if mothers had either excessive or inadequate weight gain during pregnancy. No correlation between POP exposure and gestational age was found.

Environmental Effects of POPs

POPs can travel long distances in the environment by air or in water so that they are found in ecosystems in which they have never been used and far from their source. Animals of the Canadian Arctic have significant levels of PCBs, organochlorine pesticides (like DDT), and metals, and this poses a health risk to the human population that relies on these animals as a food source. The long-range transport of POPs leads to global pollution.

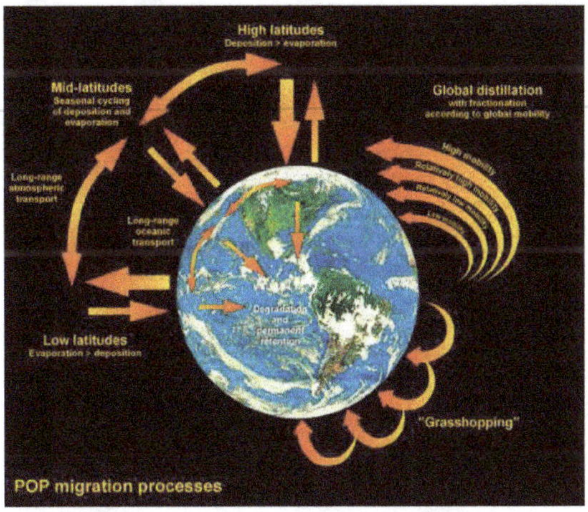

Transboundary movement of POPs.

The slow decomposition of PCDDs/PCDFs and PCBs in the environment and the hazards they pose for living organisms make them large-scale environmental degraders, especially because their toxicity can be further enhanced by their ability to accumulate in the soil and sediments and their bioaccumulation and biomagnification within aquatic and land food chains.

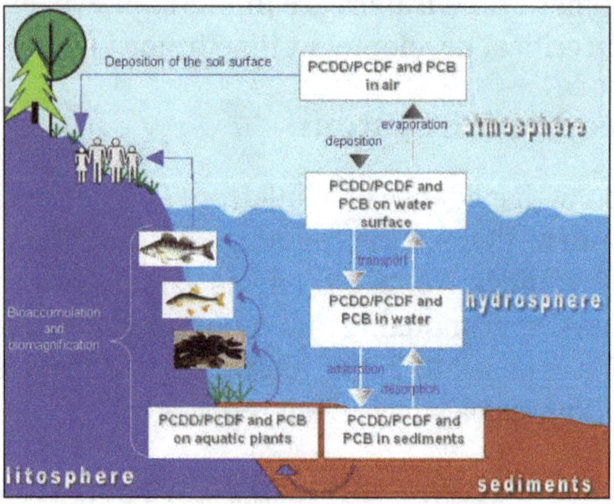

Transport and circulation of PCDDs/PCDFs and PCBs in the environment.

Degradation Pathways of POPs in the Environment

Despite the fact POPs are resistant to most of the degradation processes in the environment, some molecular alterations are possible in the environment that does not necessarily lead to simpler and less toxic compounds. Some of POP metabolites are equally complex and even more toxic than the parent molecules. Most of the degradation processes of POPs in the environment are assisted by microorganisms. Nevertheless, the half-lives of biodegradation processes of POPs are significantly long, thus accounting for their persistency in the environment.

Degradation of Heptachlor

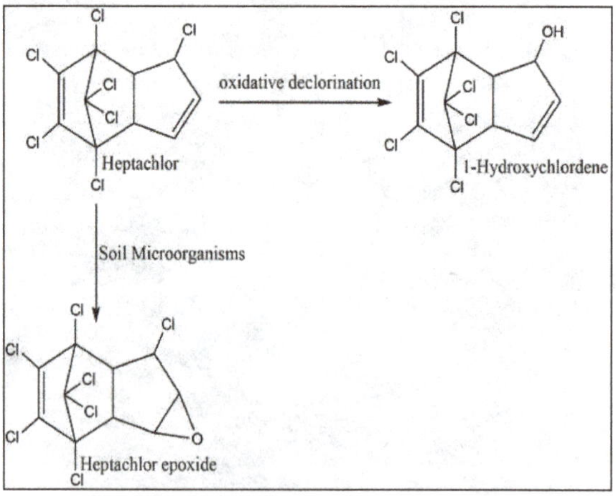

Degradation pathways of heptachlor in the soil.

Heptachlor is known to undergo oxidative dechlorinating in the soil to form 1-hydroxychlordene and by the influence of soil microorganisms to form heptachlor epoxide with a half-life of up to 112 days as shown in figure.

Degradation of DDT

p,p'-DDT can be broken down in the soil by Enterobacter aerogenes microorganisms in the presence of UV light and/or iron catalyst to reduced dehydrochlorinated compounds, DDE, and DDD as well as oxidized derivatives which ultimately form p,p'-dichlorobenzophenone. o,p'-DDT degrades in the same way as p,p'-DDT.

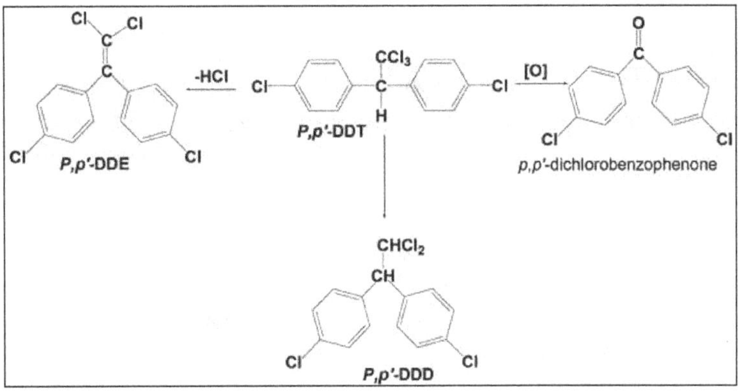

Degradation of p,p'-DDT in the environment.

Degradation of Endosulfan

The technical grade of endosulfan consists of two isomers, alpha and beta in the ratio of 70:30. In the environment, endosulfan is photolytically degraded to yield endosulfan sulfate in soil and endosulfan diol in aquatic environment. Endosulfan sulfate is equally toxic as the parent molecule.

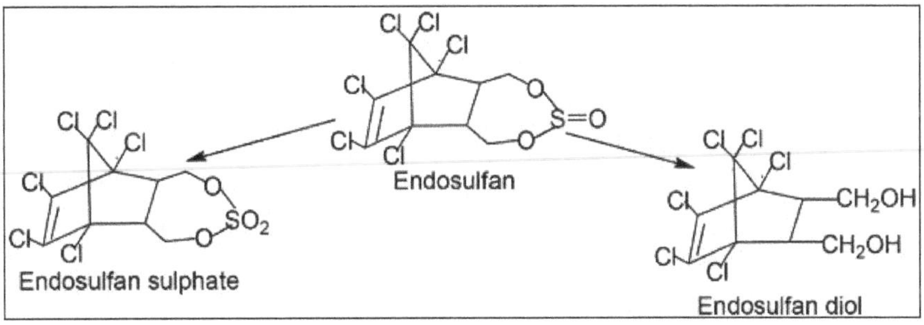

Degradation of endosulfan in the environment.

Degradation of Hexachlorobenzene

In anaerobic condition, biodegradation of HCB in an arable soil takes place with several dechlorination steps, indicating the following main HCB transformation pathways: HCB → pentachlorobenzene (QCB) → 1,2,3,5-tetrachlorobenzene (TeCB) → 1,3,5-TCB

→ 1,3-dichlorobenzene(DCB), with 1,3,5-TCB as the main intermediate dechlorination product.

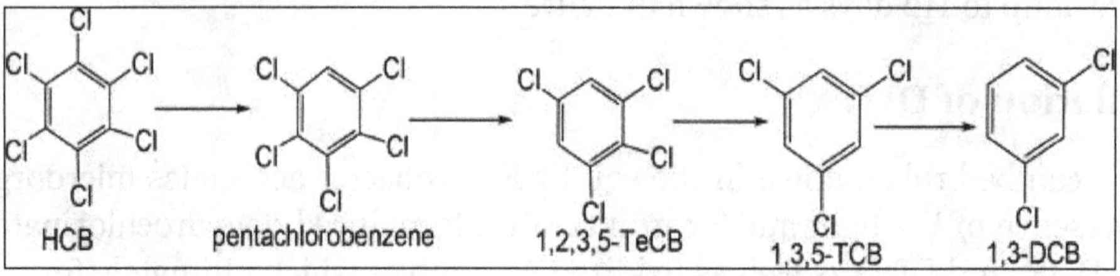

Degradation of hexachlorobenzene.

Degradation of Aldrin/dieldrin

The NADPH2-dependent enzymatic oxidation of aldrin to dieldrin and aldrin to aldrin-diol by two enzyme systems is present in excised roots. The enzymatic oxidation of aldrin in plants is known to be more species specific than the oxidation by microorganisms.

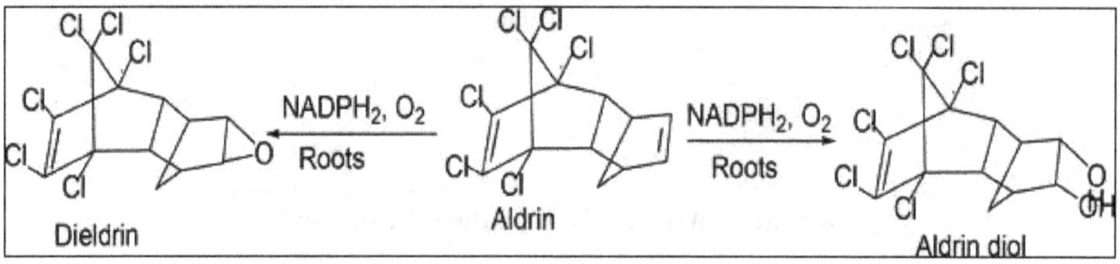

Degradation of aldrin in the environment.

Degradation of Lindane

Dehydrohalogenation of lindane to γ-hexachlorocyclohexane takes place in moist soil and is attributed to the soil microorganisms such as Bacillus coli and Clostridium sporogenes. Also these bacteria produced trace amount of benzene and monochlorobenzene from lindane.

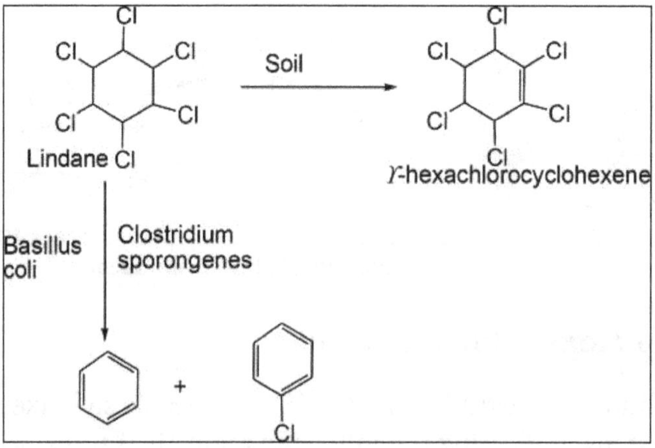

Degradation of lindane in the environment.

Degradation of PCBs

Various microorganisms are involved in a two-stage process of degradation of PCBs, which happens in aerobic and anaerobic environments. Degrading PCBs is similar to the degradation of biphenyl. However, the chlorines on PCBs prevent them from being utilized as a substrate of biphenyl degradation. Due to high chemical stability, PCBs cannot be used as energy sources. However, due to the chlorination, PCBs can be used as electron acceptors in anaerobic respiration to store energy, which is also the first stage of the degradation pathway, reductive dechlorination. Once the PCBs are dechlorinated to a certain degree, usually lower than five chlorines presenting in the structure and one aromatic ring has no chlorine, they can undergo the biphenyl degradation pathway (BP pathway) to be degraded to accessible carbon or CO_2 in the aerobic environment. BP pathway is a pathway that utilizes series of enzymes (BphA, B, C, D, E, F, G) to convert biphenyl to TCA cycle intermediates (pyruvate and Acyl-CoA) and benzoate. However, there are few microorganisms that can dechlorinate substrate under natural conditions. Even with selective media, the accumulation of PCB-dechlorinating microorganisms is still slow, which is one reason for the slow degradation rate. As a result, PCBs usually go through a co-metabolism pathway that involves different microorganism species.

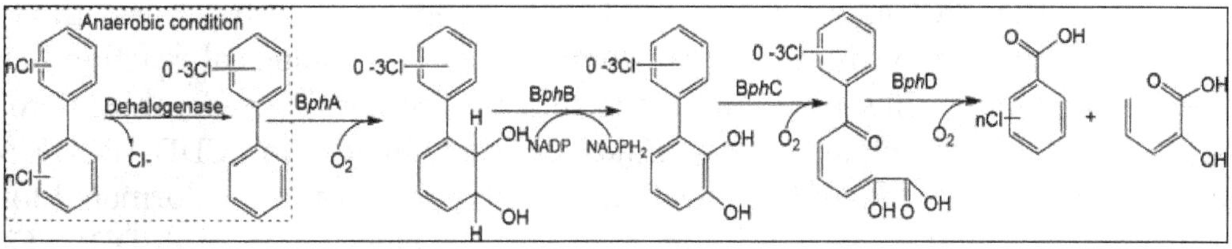

Degradation of PCBs in the environment.

Degradation of Dioxins by Reductive Dehalogenation

It has been well reported that several species of strictly or facultatively anaerobic bacteria are capable of de-halogenating chlorinated aliphatic and aromatic compounds. Some of these dehalogenation processes have been shown to couple to ATP synthesis via a chemiosmotic mechanism. The reductive dehalogenation linked to energy conservation is called "halorespiration" or "dehalorespiration." For example, a sulfate-reducing bacterium, Desulfomonile tiedjei strain DCB-1, has been shown to conserve energy for growth from reductive dehalogenation of 3-chlorobenzoate by an uncharacterized chemiosmotic process. Bacterial dehalorespiration with tri- or tetrachlorinated benzene as a terminal electron acceptor is also known to take place. Gibbs free energy of formation of various PCDDs/PCDFs and redox potentials for PCDD/PCDF substrate/product couples indicate that the reductive dehalogenation of PCDDs/PCDFs is an exergonic reaction, and this implies that microorganisms acquire energy via anaerobic electron transport with PCDDs/PCDFs as terminal electron acceptors.

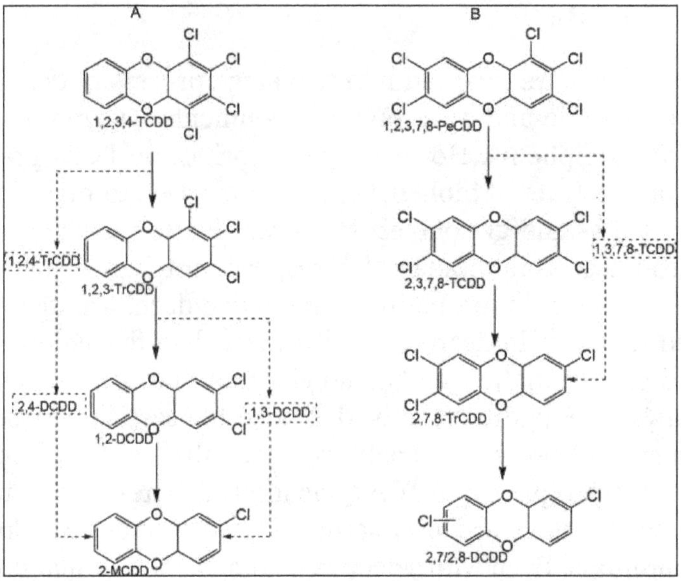

Possible degradation pathways of reductive dechlorination of 1,2,3,4-TCDD (A) and 1,2,3,7,8-PeCDD (B) by Dehalococcoides sp. strain CBDB1.

Despite its physiological and ecological importance, the biological reductive dehalogenation of PCDDs/PCDFs as well as of PCBs has been studied much less than the aerobic biodegradation of dioxin-related compounds. This biological reaction has so far been studied mostly in sediments, sludge, and soils containing anaerobic microbial consortia. Earlier work related to this subject showed changes in PCDD-/PCDF-congener distribution patterns and the resultant accumulation of less-chlorinated forms in sediments. More intensive studies on microbial dehalogenation of PCDDs/PCDFs in the environment started appearing in the past decade. Microbial dehalogenation of PCDDs/PCDFs takes place by removal mainly at positions 1, 4, 6, and/or 9 and led to much more toxic congeners, including 2,3,7,8-TCDD, in some cases. The maximum yield of cell protein coincided with the production of less-chlorinated DD isomers, where no methanogenic activity was detected. Experiments with sediment microcosms spiked with the much less toxic congener 1,2,3,4-TCDD revealed that reductive dehalogenation occurred at the lateral positions with 1,2,4-TrCDD as the main intermediate, leading to 2-MCDD as the final end product. The available information indicates that MCDDs/MCDFs are not dehalogenated further.

PERSISTENT ORGANIC POLLUTANTS IN FOOD

Persistent organic pollutants (POPs) present in foods have been a major concern for food safety due to their persistence and toxic effects. To ensure food safety and protect human health from POPs, it is critical to achieve a better understanding of POP pathways into food and develop strategies to reduce human exposure. POPs could present in food in the raw stages, transferred from the environment or artificially introduced during food preparation steps. Exposure to these pollutants may cause various health problems such as endocrine disruption, cardiovascular diseases, cancers, diabetes, birth defects, and dysfunctional immune and reproductive systems.

Chemical contaminants have become a concern in terms of food safety due to pesticide residue and environmental contaminants detected in the food supply. A large amount of pollutants generated from rapidly developing agricultural and industrial sectors have been released to the environment and found their way into the food supply. Given the widespread occurrence of chemical contaminants in foodstuff and their serious health risks, the prevention of food contamination is a public health priority. In recent decades, there has been a focus on a subset of harmful organic chemicals, mostly of anthropogenic origin, that are commonly classified as persistent organic pollutants (POPs). POPs are a class of carbon-based organic chemicals that are persistent, bioaccumulative and have long-range transport potential.

There are three types of POPs present in the environment:

- Pesticides, especially organochlorine pesticides (OCPs) such as dichlorodiphenyltrichloroethane (DDT) and its metabolites.
- Industrial and technical chemicals including polychlorinated biphenyls (PCBs), polybrominated diphenyl ethers (PBDEs), and perfluorooctanesulfonate (PFOS).
- By-products of industrial processes including polychlorinated dibenzo-p-dioxins (PCDDs), polychlorinated dibenzofurans (PCDFs), and polyaromatic hydrocarbons (PAHs).

PAHs do not strictly belong to POPs and they are only recognized as POPs under the Aarhus Protocol because they can be efficiently metabolized and, therefore, prevent further bioaccumulation. However, due to their lipophilicity and continuous release, PAHs are frequently classified as POPs in many studies.

Most POPs are halogenated chemicals and the strong bond between carbon and chlorine/bromine/fluorine makes POPs resistant to the environmental degradation including chemical, biological, and photolytic reactions. For those non-halogenated POPs, their stable chemical structures also make them persistent in nature. POPs are resistant to environmental degradation including chemical, biological, and photolytic reactions.

Therefore, once released, POPs can stay in the environment for a long time. Some POPs could have half-lives of years or decades and they can stay in the environment until they are taken up by plants and animals. POPs can bioaccumulate in the fatty tissue of living organisms and, therefore, become concentrated as they move through the food chain.

POPs are a small subset of persistent, bioaccumulative and toxic chemicals (PBTs) that can travel great distance. With the POPs group, some POPs are easily transportable, and others are not. POPs can move long distances in the atmosphere through air and water, even to places where they have never been used such as Antarctica and the Arctic area. Due to these features, human and animals around the world could be exposed to POPs for extended long periods of time. Exposure to these pollutants, mostly at high levels, may cause various health problems such as endocrine disruption, cardiovascular diseases, cancers, diabetes, birth defects, and dysfunctional immune and reproductive systems. The peak release of POPs was in the 1970s. Due to effective regulation and legislation, the current concentrations of many POPs are only one-tenth of the concentration at that time. In developed countries, many POPs are monitored to be below safety limits based on the known toxicology information. POPs are more of a threat to humans historically than now. However, POPs remain as a concern to human health because of the chronic exposure and the accumulation of POPs in the human body, especially in some developing countries.

Over 90% of human exposure to POPs is through the consumption of contaminated food, particularly food of animal origin. Fishes are among the major sources of exposure to POPs. To better protect public health, it is important to understand POP pathways into food, and the environment is among the significant pathways. POPs have been used and released to the environment through various human activities such as through the industrial and agriculture sector. The released POPs can easily contaminate crops, livestock, seafood and drinking water and pose a high risk to human health. For example, pesticides such as DDT and dieldrin have been widely used in the agriculture to increase crop yield and to kill unwanted pests in recent decades. However, the application of OCPs can easily introduce contaminants into water and crops and wildlife. Studies find that pesticide residue is among the most commonly found food contaminants. Similarly, other POPs present in the environment such as PCBs, hexachlorobenzenes (HCBs), dioxin, and furans are also commonly found food contaminants.

To protect consumers from POP-contaminated food, many national and international agencies such as the European Food Safety Authority (EFSA), World Health Organization (WHO), US Environmental Protection Agency (EPA), and US Food and Drug Administration (FDA) have developed regulations and guidelines to reduce the exposure to POPs. The Stockholm Convention requires its parties to take action to decrease the production, use and releases of the POPs on its list. The initial list was established in 2001 and included 12 POPs that are known as the "dirty dozen". By 2019, an additional 17 POPs were added to the list. The Stockholm Convention is the most prominent, legal binding international framework that prevents the further accumulation of persistent toxic chemicals in food at the global level. As a result of the regulations and legal frameworks, the emission of dioxin and dioxin-like compounds have decreased considerably, as well as some hazardous pesticides in recent decades. Some monitoring programs have been employed to trace the trend of POPs such as the global monitoring plan for

POPs by the United Nations Environment Programme (UNEP). The surveys show a consistent decline in PCDD, PCDF, PCB and DDT levels in human milk. Overall, a general decrease was observed for POP levels in the environment and population in recent decades, demonstrating the effectiveness of the regulations and legal frameworks.

Currently, one challenging task in food safety is the assessment of health risks associated with POP dietary exposure. To assess the risk, information on the toxic effects of POPs and levels of dietary exposure is needed. However, for most POPs, the toxic information is hardly available, mostly from quantitative structure activity relationship (QSAR) models, animal experiments, and epidemiological studies. For POPs with quantified acute toxic effects, the estimation of their health risks is rather straightforward, and authorities could protect public health by limiting these POPs to below the safety limit. However, for most POPs found in food, the effects are long term and chronic. The continuous exposure and accumulation of POPs in the human body make it difficult to assess the dose–response relation between POPs and potential health problems, let alone determine the effective dose. For some POPs such as dioxin, furan, dioxin-like PCBs, PFOA and PFOS, some authorities have established the threshold values based on known toxicological information, but for a number of other commonly found POPs, the toxicological information is not clear, and the safety level is still unknown. Moreover, it is impossible to determine the threshold for genotoxic carcinogens such as some PAHs and their epoxides, because these POPs could damage DNA and there is no safety level of exposure. For those POPs, a margin of exposure approach is used to determine the level of risks. The European Commission has established the maximum levels of certain PAHS on different foodstuff.

Dietary intake of POPs is usually estimated through the total diet study in various countries and high performance analytical methods are needed in the study to determine POP levels in food. Currently, effective and rapid analytical approaches have been developed to determine trace amounts of POPs in food. The approaches usually require multistep strategies including sample preparation and highly selective and sensitive instrumental techniques under strict quality assurance/quality control criteria. The choice of sample preparation technique depends on the characteristics of the matrix. Sample preparation may include filtration, pH adjustment, extraction, clean-up and preconcentration procedures to ensure that the analytes are found at a suitable concentration level. Popular sample extraction methods include Soxhlet extraction (SOX), solid–liquid extraction (SLE), solid-phase extraction (SPE), solid-phase microextraction (SPME), microwave-assisted extraction (MAE), liquid–liquid extraction (LLE), pressurized liquid extraction (PLE) and stir bar sorptive extraction (SBSE).

Mass spectrometry (MS) has been considered among the most suitable instruments for the detection of POPs in food. MS has been widely applied in analyses of POPs in food due to its advantage of high sensitivity, selectivity and throughput. MS coupling with suitable separation techniques such as liquid chromatography (LC–MS), gas chromatography (GC–MS) and two-dimensional gas chromatography (GC×GC–MS) has been widely used for the detection of POPs in food due to its easy automation and high speed.

Sources of POP Contamination in Food

Food preparation usually involves multiple steps including processing, packaging, transportation and storage. Each step could be a potential source for POP invasion. Food could be contaminated by POPs through different paths. For example, raw materials may contain POPs that are transferred from the environment. Since POPs are resistant to degradation, they can stay in the environment for an extended period time. Previously released POPs in the environment are a major source of the POP contamination of food and feed supplies. Plant foliage uptake of POPs can effectively transfer POPs from air to plant and subsequently to food. Other sources of POPs are food preparation steps, during which POPs may artificially be introduced by humans.

OCPs

OCPs are among the main classes of POPs in the environment. They have been widely used in the agriculture industry since the second world war to protect plants. With a low cost, high toxicity and persistence nature, OCPs make the ideal candidates for treating soil and plants against various insects. However, OCPs could cause serious health problems for humans. The dietary exposure of OCPs is mainly from contaminated fatty foods including foods of animal origin such as eggs, dairy product and meat and foods of plant origin such as rice, fruit, vegetables and vegetable oil. Animals could be contaminated if they have access to OCP residues or OCP-contaminated feeds. Vegetables may be contaminated by root uptake from contaminated soil or from direct contact with OCPs.

Due to persistent toxicity, many OCPs such as aldrin, chlordane, DDT, dieldrin, endrin, heptachlor, HCB, mirex and toxaphene have been banned or restricted for use. However, these OCP residues remain in the environment due to their persistent nature. In addition, some OCPs are still used in areas such as Africa, South Asia, Central and South America. For example, high HCB levels were found in oysters from the pacific coast of Mexico. A possible explanation for this is that the coast is bordered by extensive agriculture lands and HCBs were used in the Mexico's agriculture until recently. Barber et al. confirmed the observation by reviewing the distribution and levels of HCB in the global environment. Another example is DDT, which was banned from agriculture use in most industrialized countries in the 1970s and 1980s because of its persistence and toxic effects. However, many developing countries still use DDT in controlling the spread of malaria. Since this public health benefit outweighs the potential adverse effects it may cause, the Stockholm Convention lists DDT in its annex B, allowing the production and use of DDT for the purpose of disease vector control. High concentrations of DDT were found in chickens and chicken eggs in malaria-controlled areas of South Africa where

DDT was used as an indoor residual spray. Levels of DDT and HCB were also detected in fish samples because of the elevated use of pesticides in agriculture until recently and the continuous use of DDT in controlling the spread of malaria.

PCB/BDE

PCB and PBDE are two POPs that have a similar structure and similar physicochemical properties. PCBs were widely used in the 20th century as coolants and lubricants in transformers and capacitors and as hydraulic and heat exchange fluids in electrical apparatus. PBDEs were extensively used from the beginning of the 1960s as flame retardants in electrical appliances, computers, TVs, insulation wires, cables and building materials. The waste of electrical and electronic equipment has been released to the environment and resulted in high concentrations of PCBs and PBDEs in fatty foods such as meat, fish and poultry. Humans could be exposed to PCB/BDE through inhalation, dermal contact and the consumption of contaminated food. Dietary exposure is the main route for PCB/BDE accumulation in the human body. For PCBs, there are 209 possible congeners and analysis is usually focused on toxic PCBs including dioxin-like PCBs (PCB-77, 81, 105, 114, 118, 123, 126, 156, 157, 167, 169, and 189) and indicator PCBs (PCB-28, 52, 101, 138, 153 and 180). For PBDEs, the most detected congeners in foodstuffs include 16 PBDEs (BDE-28, 47, 49, 53, 66, 85, 99, 100, 153, 154, 183, 196, 206, 207, 208 and 209).

Although PCBs were banned in the 1970s, the old electrical transformers and building materials built before the ban still release PCBs into the environment and this is more significant than the incineration of PCB-containing waste and vehicle emission. In addition, previously released PCBs persist in the environment and food sources. Over 90% of the PCB body burden comes from food intake, mainly of fish, meat and dairy products. Son et al. studied the distribution of PCBs in 40 different foodstuffs in Korea and found that fish had approximately 4-fold higher concentrations of PCBs than rice. However, even though foodstuff such as vegetables and rice have a low concentration of PCBs, they should be considered a potential and significant source of PCB intake because of the large amount consumed. Dietary makeup is a very important factor in the risk assessment of POP exposure. PBDE was banned in the USA and Europe in 2004. However, products manufactured with PBDE are still in use, which could be the major source of PBDE contamination. PCB/BDE-contaminated food can be found in countries where these chemicals have never been produced because of imported equipment containing PCB/BDE due to their long-range transportability.

Dioxin and Furan

PCDD and PCDF are known by-products of many industrial processes involving chlorinated chemical compounds. Humans are exposed to dioxins/furans mainly through high-fat food such as fish, meat and dairy products. The waste incineration is a major source of dioxins and furans in developing countries where medical waste incineration and open burning are common. Therefore, high levels of dioxins/furans were found in the milk and meat of animals in the vicinity of incineration plants. For example, high levels of dioxins/furans and dioxin-like PCBs were found in some food groups (rice, vegetables, chicken, hen eggs, duck and crucian carp) from Zhejiang province, China, due to uncontrolled electronic recycling operations.

Accidents caused by artificially introduced POPs were well documented such as Yusho poisoning in Japan 1968 and Yu-cheng poisoning in Taiwan in 1979. In the Yusho/Yucheng accidents, the victims were exposed to rice oil contaminated by extremely high concentrations of PCDFs/PCBs. The toxic equivalent (TEQ) intake for the accident victims was almost four-fold higher than for the general population. The exposure continued over an extended period and caused serious health problems. On the other hand, for some other food poisoning accidents, such as the Belgian dioxin crisis in 1999 and the 2008 dioxin accident in Ireland, the dioxin burden on the population did not increase and no consequences have been reported. Because slow elimination leads to slow accumulation, the body burden depends more on long-term exposure than occasional exceedances of tolerable daily intake limits. The Yusho and Yucheng accidents are different due to the extremely high levels of contaminants involved. There are other unintentional contaminated feed supply examples such as ball clays in poultry, kaolinic clay in Netherlands and catfish feeds. Accidental POP food and feed contamination is among the primary sources of single, extremely high dietary exposure to POPs. These incidents have aroused public attention to POP-related food safety issues. National and international agencies have developed policies to reduce consumer exposure to POPs by setting food and feed limits and effective surveillance programs to regulate POP levels in food and feed supply.

PAHs and Per- and Polyfluoroalkyl Substances

PAHs are a ubiquitous group of several hundreds of chemicals that comprise two or more benzene rings. PAHs are generated mostly because of the incomplete combustion of either natural sources such as coal and wood or man-made sources such as automobile emissions and cigarette smoke. For non-smokers, diet exposure contributes over 70% of the human exposure. PAHs may enter food through food preparation and handling methods. A dietary survey in the United Kingdom showed that cereals and oils/fats contributed a large part to the dietary intake of PAHs. This typical PAH contamination usually occurs in technological processes such as direct fire drying, where food is exposed to combustion products. Certain traditional home cooking methods such as grilling, roasting, frying and smoking could also result in high PAH concentrations in charcoal grilled/barbecued foods. The packing materials and manufacturing are another source of potential food contamination.

Per- and polyfluoroalkyl substances (PFASs) and their derivatives are a group of chemicals comprising a fluorinated alkyl chain (generally C4–C18) and a hydrophilic functional group. Since these chemicals are waterproof, oil-proof and heat resistant, they are used in a wide range of products and applications such as food packaging, non-stick cookware, cleaning agents, carpet, furniture, coating materials, etc.. The most investigated PFASs include perfluorooctanesulfonate (PFOS) and perfluorooctanoic acid (PFOA). Perfluorooctanesulfonate is used in paper coating for food contact and perfluorooctanoic acid is a processing aid used in nonstick cookware. PFOS has been detected in many food sources. The main sources of PFOAs and PFOSs are fish, seafood, meat, eggs, dairy product and drinking water. Recent studies on fluorinated chemicals in food packaging show their potentially significant contribution to dietary exposure to PFASs.

References

- Organic-pollutant, medicine-and-dentistry: sciencedirect.com, Retrieved 10 July, 2020
- Organic-pollutants, pollution: worldoceanreview.com, Retrieved 11 May, 2020
- Persistent-organic-pollutants-general-considerations-316623280: researchgate.net, Retrieved 31 March, 2020
- Stockholm-Convention-on-Persistent-Organic-Pollutants-POPs, Convention: chemsafetypro.com, Retrieved 21 August, 2020
- Degradation-pathways-of-persistent-organic-pollutants-pops-in-the-environment, persistent-organic-pollutants: intechopen.com, Retrieved 06 April, 2020

Types of Organic Pollutants

CHAPTER 2

There are many types of organic pollutants which include aldrin, chlordane, dieldrin, endrin, polychlorinated biphenyl, dioxins, pentachlorobenzene, endosulfan, organochloride, etc. The topics elaborated in this chapter will help in gaining a better perspective of the different types of organic pollutants.

ALDRIN

- Aldrin is a pesticide used to control soil insects such as termites, corn rootworm, wireworms, rice water weevil, and grasshoppers. It has been widely used to protect crops such as corn and potatoes, and has been effective to protect wooden structures from termites.
- Aldrin is readily metabolized to dieldrin by both plants and animals. As a result, Aldrin residues are rarely found in foods and animals, and then only in small amounts. It binds strongly to soil particles and is very resistant to leaching into groundwater. Volatilization is an important mechanism of loss from soils. Due to its persistent nature and hydrophobicity, Aldrin is known to bioconcentrate, mainly as its conversion products.
- Aldrin is banned in many countries, including Bulgaria, Ecuador, Finland, Hungary, Israel, Singapore, Switzerland, and Turkey. Its use is severely restricted in many countries, including Argentina, Austria, Canada, Chile, the EU, Japan, New Zealand, the Philippines, USA, and Venezuela.

Aldrin Container.

Effects on Humans

Aldrin is toxic to humans; the lethal dose of Aldrin for an adult man has been estimated to be about 5g, equivalent to 83 mg/kg body weight. Signs and symptoms of Aldrin intoxication may include headache, dizziness, nausea, general malaise and vomiting, followed by muscle twitching, myoclonic jerks, and convulsions. Occupational exposure to Aldrin, in conjunction with Dieldrin and Endrin, was associated with a significant increase in liver and biliary cancer, although the study did have some limitations, including a lack of quantitative exposure information. There is limited information that cyclodienes, such as Aldrin, may affect immune responses.

The acute oral LD50 dose for Aldrin in laboratory animals is in the range of 33 mg/kg body weight for guinea pigs to 320 mg/kg body weight for hamsters. Reproductive effects in rats were observed when pregnant females were dosed with 1.0 mg/kg Aldrin subcutaneously. Offspring experienced a decrease in the median effective time for incisor teeth eruption and an increase in the median effective time for testes descent. There is, as yet, no evidence of a teratogenic potential for Aldrin. The International Agency for Research on Cancer (IARC) has concluded that there is inadequate evidence for the carcinogenicity of Aldrin in humans, and there is only limited evidence in experimental animals. Aldrin is therefore not classifiable as to its carcinogenicity in humans.

The average daily human intake of Aldrin and Dieldrin was calculated to be 19μg/person in India, and 0.55 μg/person in Vietnam. Dairy products, such as milk and butter, and animal meats are the primary sources of exposure.

Effects on Plants and Animals

Aldrin has low phytotoxicity with plants affected only by extremely high application rates. The toxicity of Aldrin to aquatic organisms is quite variable, with aquatic insects being the most sensitive group of invertebrates. The 96-h LC50 dose values range from 1-200 μg/L for insects, and from 2.2-53 μg/L for fish. Long term and bioconcentration studies are performed primarily using Dieldrin, the primary conversion product of Aldrin. In a model ecosystem study, only 0.5% of the original radioactive Aldrin was stored as Aldrin in the mosquitofish (Gambusia affinis), the organism at the top of the model food chain.

The acute toxicity of Aldrin to avian species varies in the range of 6.6 mg/kg for bobwhite quail to 520 mg/kg for mallard ducks. Aldrin treated rice is thought to have been the cause of deaths in waterfowl, shorebirds and passerines along the Texas Gulf Coast, both by direct poisoning by ingestion of Aldrin treated rice and indirectly by consuming organisms contaminated with Aldrin. Residues of Aldrin were detected in all samples of bird casualties, eggs, scavengers, predators, fish, frogs, invertebrates and soil.

As Aldrin is readily and rapidly converted to Dieldrin in the environment its fate is closely linked to that of Dieldrin. Aldrin is readily metabolised to dieldrin in both animals and plants, and therefore Aldrin residues are rarely present in animals and then only in very small amounts. Residues of Aldrin have been detected in fish in Egypt, the average concentration was 8.8 µg/kg, and a maximum concentration of 54.27 µg/kg.

Structure and Reactivity

The structure formula of aldrin is $C_{12}H_8Cl_6$. The molecule has a molecular weight of 364.896 g/mol. The melting point of aldrin is a temperature of 105 °C and the octanol-water partition coefficient is 6.5 (logP).

Pure aldrin takes form as a white crystalline powder. Though it is not soluble in water (0.003%% solubility), aldrin dissolves very well in organic solvents, such as ketones and paraffins. Aldrin decays very slowly once released into the environment. Though it is rapidly converted to dieldrin by plants and bacteria, dieldrin maintains the same toxic effects and slow decay of aldrin. Aldrin is easily transported through the air by dust particles. Aldrin does not react with mild acids or bases and is stable in an environment with a pH between 4 and 8. It is highly flammable when exposed to temperatures above 200 °C In the presence of oxidizing agents aldrin reacts with concentrated acids and phenols.

Synthesis

Aldrin is not formed in nature. It is synthesized by combining hexachlorocyclopentadiene with norbornadiene in a Diels-Alder reaction to give the adduct. In 1967, the composition of technical-grade aldrin was reported to consist of 90.5% of hexachlorohexahydrodimethanonaphthalene (HHDN).

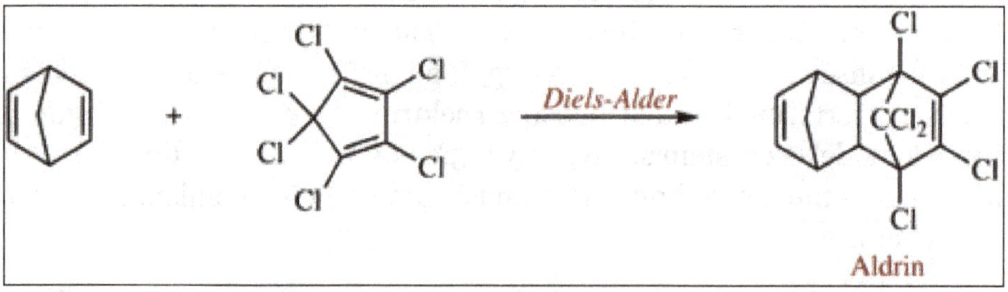

Synthesis of aldrin via a Diels-Alder reaction.

Similarly, an isomer of aldrin, known as isodrin, is produced by reaction of hexachloronobornadiene with cyclopentadiene. Isodrin is also produced as a byproduct of aldrin synthesis, with technical-grade aldrin containing about 3,5% isodrin.

Aldrin is named after the German chemist Kurt Alder, one of the coinventors of this kind of reaction. An estimated 270 million kilograms of aldrin and related cyclodiene pesticides were produced between 1946 and 1976.

Available Forms

There are multiple available forms of aldrin. One of these is the isomer isodrin, which cannot be found in nature, but needs to be synthesized like aldrin. When aldrin enters the human body or the environment it is rapidly converted to dieldrin. Degradation by ultraviolet radiation or microbes can convert dieldrin to photodieldrin and aldrin to photoaldrin.

Mechanism of Action

Even though many toxic effects of aldrin have been discovered, the exact mechanisms underlying the toxicity are yet to be determined. The only toxic aldrin induced process that is largely understood is that of neurotoxicity.

Neurotoxicity

One of the effects that intoxication with aldrin gives rise to is neurotoxicity. Studies have shown that aldrin stimulates the central nervous system (CNS), which may cause hyperexcitation and seizures. This phenomenon exerts its effect through two different mechanisms. One of the mechanisms uses the ability of aldrin to inhibit brain calcium ATPases. These ion pumps release the nerve terminal from calcium by actively pumping it out. However, when aldrin inhibits these pumps, the intracellular calcium levels rise. This results in an enhanced neurotransmitter release.

The second mechanism makes use of aldrin's ability to block gamma-aminobutyric acid (GABA) activity. GABA is a major inhibitory neurotransmitter in the central nervous system. Aldrin induces neurotoxic effects by blocking the $GABA_A$ receptor-chloride channel complex. By blocking this receptor, chloride is unable to move into the synapse, which prevents hyperpolarization of neuronal synapses. Therefore, the synapses are more likely to generate action potentials.

Metabolism

The metabolism of oral aldrin exposure has not been studied in humans. However, animal studies are able to provide an extensive overview of the metabolism of aldrin. This data can be related to humans.

Biotransformation of aldrin starts with epoxidation of aldrin by mixed-function oxidases (CYP-450), which forms dieldrin. This conversion happens mainly in the liver. Tissues with low CYP-450 expression use the prostaglandin endoperoxide synthase (PES) instead. This oxidative pathway bisdioxygenises the arachidonic acid to prostaglandin G_2 (PGG_2). Subsequently, PGG_2 is reduced to prostaglandin H_2 (PGH_2) by hydroperoxidase.

Dieldrin can then be directly oxidized by cytochrome oxidases, which forms

9-hydroxydieldrin. An alternative for oxidation involves the opening of the epoxied ring by epoxied hydrases, which forms the product 6,7-trans-dihydroxydihydroaldrin. Both products can be conjugated to form 6,7-trans-dihydroxydihydroaldrin glucuronide and 9-hydroxydieldrin glucuronide, respectively. 6,7-trans-dihydroxydihydroaldrin can also be oxidized to form aldrin dicarboxylic acid.

Efficacy and Side Effects

Considering the toxicokinetics of aldrin in the environment, the efficacy of the compound has been determined. In addition, the adverse effects after exposure to the aldrin are demonstrated, indicating the risk regarding the compound.

- Efficacy: The ability of aldrin, in its use for the control of termites, is examined in order to determine the maximum response when applied. In 1953 US researchers tested aldrin and dieldrin on terrains with rats known to carry chiggers, at a rate of 2.25 pound per acre. The Aldrin and Dieldrin treatment demonstrated a decrease of 75 times less chiggers on rats for Dieldrin treated terrains and 25 times less chiggers on the rats when treated with Aldrin. The Aldrin treatment indicate a high productivity, especially in comparison to other insecticide that were used, like DDT, Sulfur or BHC.
- Adverse effects: Exposure of Aldrin to the environment leads to the localization of the chemical compound in the air, soil, and water. Aldrin gets changed quickly to dieldrin and that compound degrades slowly, which accounts for aldrin concentrations in the environment around the primary exposure and in the plants. These concentrations can also be found in animals, which eat contaminated plants or animals that reside in the contaminated water. This biomagnification can lead to a high concentrations in their fat.

There are some reported cases of workers who developed anemia after multiple dieldrin exposures. However the main adverse effect of Aldrin and Dieldrin is in relationship to the central nervous system. The accumulated levels of Dieldrin in the body were believed to lead to convulsions. Besides that other symptoms were also reported like headaches, nausea and vomiting, anorexia, muscle twitching and myoclonic jerking and EEG distortions. In all these cases removal of the source of exposure to aldrin/dieldrin led to a rapid recovery.

Toxicity

The toxicity of aldrin and dieldrin is determined by the results of several animal studies. Reports of a significant increase in workers death in relation to aldrin has not been found, although death by anemia is reported in some cases after multiple exposure to Aldrin. Immunological tests linked an antigenic response to erythrocytes coated with dieldrin in those cases. Direct dose-response relations being a cause for death are yet to be examined.

The NOAEL that was derived from rat studies:

- The minimal risk level at acute oral exposure to Aldrin is 0.002 mg/kg/day.
- The minimal risk level at intermediate exposure to Dieldrin is 0.0001 mg/kg/day.
- The minimal risk level at chronic exposure to Aldrin is 0.00003 mg/kg/day.
- The minimal risk level at chronic exposure to Dieldrin is 0.00005 mg/kg/day.

In addition to these studies, breast cancer risk studies were performed demonstrating a significant increased breast cancer risk. After comparing blood concentrations to number of lymph nodes and tumor size a 5-fold higher risk of death was determined, comparing the highest quartile range in the research to the lower quartile range. Young children are also more susceptible to the drug, causing severe generalized convulsions.

Safety and Environmental Aspects

Aldrin has rat LD_{50} of 39 to 60 mg/kg (oral in rats). For fish however, it is extremely toxic, with an LC50 of 0.006 – 0.01 for trout and bluegill. In the US, aldrin is considered a potential occupational carcinogen by the Occupational Safety and Health Administration and the National Institute for Occupational Safety and Health; these agencies have set an occupational exposure limit for dermal exposures at 0.25 mg/m^3 over an eight-hour time-weighted average. Further, an IDLH limit has been set at 25 mg/m^3, based on acute toxicity data in humans to which subjects reacted with convulsions within 20 minutes of exposure.

It is classified as an extremely hazardous substance in the United States as defined in Section 302 of the U.S. Emergency Planning and Community Right-to-Know Act (42 U.S.C. 11002), and is subject to strict reporting requirements by facilities which produce, store, or use it in significant quantities.

CHLORDANE

Chlordane, or chlordan, is an organochlorine compound used as a pesticide. It is a white solid. In the United States, chlordane was used for termite-treatment of approximately 30 million homes until it was banned in 1988. Chlordane was banned 10 years earlier for food crops like corn and citrus, and on lawns and domestic gardens.

Like other chlorinated cyclodiene insecticides, chlordane is classified as an organic pollutant hazardous for human health. It is resistant to degradation in the environment and in humans/animals and readily accumulates in lipids (fats) of humans and animals. Exposure to the compound has been linked to cancers, diabetes, and neurological disorders.

Production, Composition and Uses

Technical chlordane development was by chance by Julius Hyman in 1948, during a search for possible uses of a by-product of synthetic rubber manufacturing. By chlorinating this by-product, persistent and potent insecticides were easily and cheaply produced. The chlorine atoms, 7 in the case of heptachlor and 8 in chlordane, and 9 in the case of nonachlor, surround and stabilize the cyclodiene ring and thus these compounds are referred to as cyclodienes. Other members of the cyclodiene family of organochlorine insecticides are aldrin and its epoxide, dieldrin, as well as endrin, which is a stereoisomer of dieldrin. Cyclodiene derives its name from hexachlorocyclopentadiene a precursor in its production.

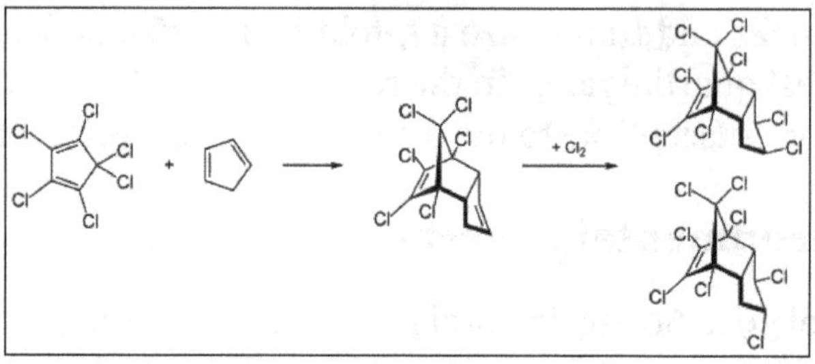

Synthesis of cis- (above) and trans-chlordane (below).

Hexachlorocyclopentadiene forms a Diels-Alder adduct with cyclopentadiene to give Chlordene intermediate [3734-48-3]; chlorination of this adduct gives predominantly two chlordane isomers, α and β, in addition to other products such as *trans*-nonachlor and heptachlor. The β-isomer is popularly known as gamma and is more bioactive. The mixture that is composed of 147 components is called technical chlordane.

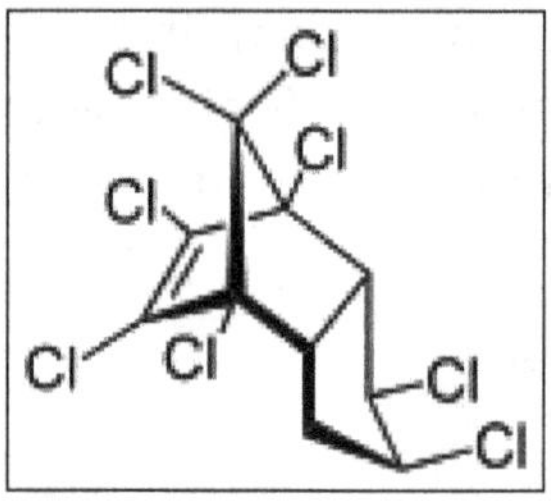

Cis-chlordane (also known as α-chlordane (CAS=5103-71-9)).

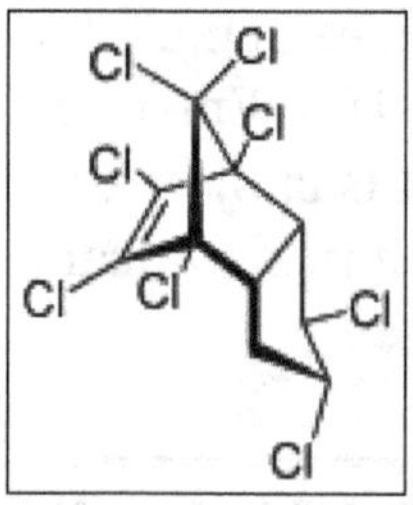

Trans-chlordane (also known as γ-chlordane and beta-chlordane (CAS=5103-74-2)).

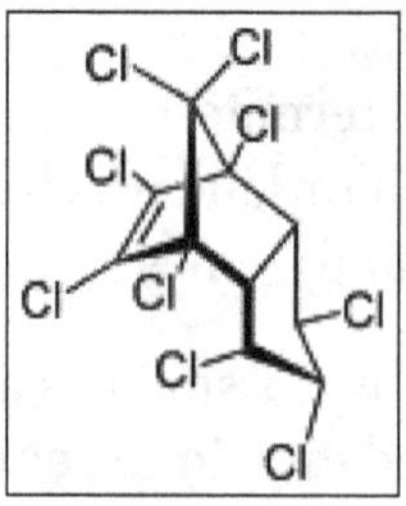

Trans-nonachlor.

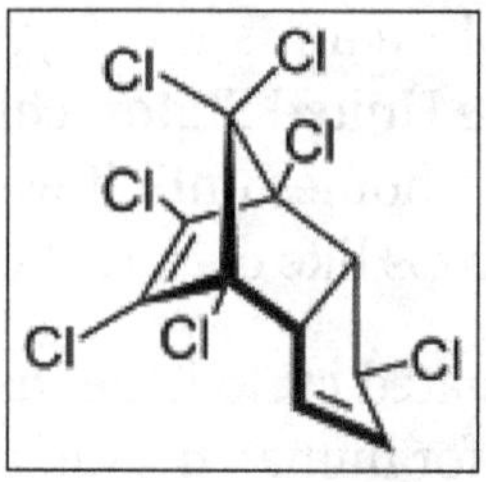

(+)-heptachlor.

Chlordane appears as a white or off-white crystals when synthesized, but it was more commonly sold in various formulations as oil solutions, emulsions, sprays, dusts, and powders. These products were sold in the United States from 1948 to 1988.

Because of concern for harm to human health and to the environment, the United States Environmental Protection Agency (EPA) banned all uses of chlordane in 1983, except termite control in wooden structures (e.g. houses). After many reports of chlordane in the indoor air of treated homes, EPA banned the remaining use of chlordane in 1988. The EPA recommends that children should not drink water with more than 60 parts of chlordane per billion parts of drinking water (60 ppb) for longer than 1 day. EPA has set a limit in drinking water of 2 ppb.

Chlordane is very persistent in the environment because it does not break down easily. Tests of the air in the residence of U.S. government housing, 32 years after chlordane treatment, showed levels of chlordane and heptachlor 10-15 times the Minimal Risk Levels (20 nanograms/cubic meter of air) published by the Centers for Disease Control. It has an environmental half-life of 10 to 20 years.

Origin, Pathways of Exposure and Processes of Excretion

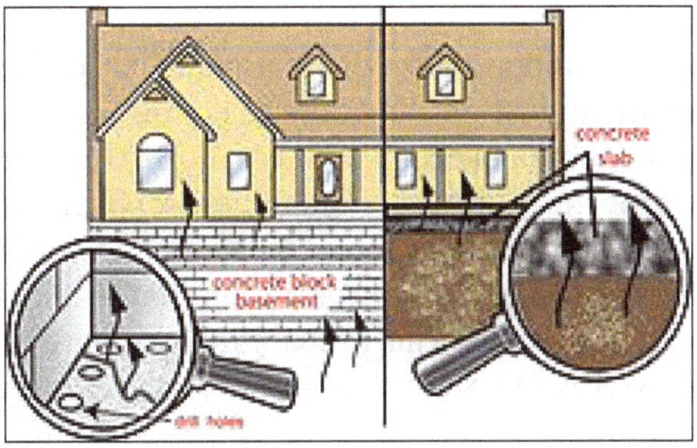

Sources and pathways that chlordane contaminates the indoor air of American homes.

In the years 1948–1988 chlordane was a common pesticide for corn and citrus crops, as well as a method of home termite control. Pathways of exposure to chlordane include ingestion of crops grown in chlordane-contaminated soil, inhalation of air in chlordane-treated homes and from landfills, and ingestion of high-fat foods such as meat, fish, and dairy, as chlordane builds up in fatty tissue. The United States Environmental Protection Agency reported that over 30 million homes were treated with technical chlordane or technical chlordane with heptachlor. Depending on the site of home treatment, the indoor air levels of chlordane can still exceed the Minimal Risks Levels (MRLs) for both cancer and chronic disease by orders of magnitude. Chlordane is excreted slowly through feces, urine elimination, and through breast milk in nursing mothers. It is able to cross the placenta and become absorbed by developing fetuses in pregnant women. A breakdown product of chlordane, the metabolite oxychlordane, accumulates in blood and adipose tissue with age.

Environmental Impact

Being hydrophobic, chlordane adheres to soil particles and enters groundwater only slowly, owing to its low solubility (0.009 ppm). It requires many years to degrade. Chlordane bioaccumulates in animals. It is highly toxic to fish, with an LD_{50} of 0.022–0.095 mg/kg (oral).

Oxychlordane ($C_{10}H_4Cl_8O$), the primary metabolite of chlordane, and heptachlor epoxide, the primary metabolite of heptachlor, along with the two other main components of the chlordane mixture, *cis*-nonachlor and *trans*-nonachlor, are the main bioaccumulating constituents. *trans*-Nonachlor is more toxic than technical chlordane and *cis*-nonachlor is less toxic. Chlordane and heptachlor are known as persistent organic pollutants (POP), classified among the "dirty dozen" and banned by the 2001 Stockholm Convention on Persistent Organic Pollutants.

Health Effects

Exposure to chlordane/heptachlor and/or its metabolites (oxychlordane, heptachlor epoxide) are risk factors for type-2 diabetes, for lymphoma, for prostate cancer, for obesity, for testicular cancer, for breast cancer.

An epidemiological study conducted by the National Cancer Institute reported that higher levels of chlordane in dust on the floors of homes were associated with higher rates of non-Hodgkin lymphoma in occupants. Breathing chlordane in indoor air is the main route of exposure for these levels in human tissues. Currently, EPA has defined a concentration of 24 nanogram per cubic meter of air (ng/M^3) for chlordane compounds over a 20-year exposure period as the concentration that will increase the probability of cancer by 1 in 1,000,000 persons. This probability of developing cancer increases to 10 in 1,000,000 persons with an exposure of 100 ng/m^3 and 100 in 1,000,000 with an exposure of 1000 ng/m^3.

The non-cancer health effects of chlordane compounds, which include diabetes, insulin resistance, migraines, respiratory infections, immune-system activation, anxiety, depression, blurry vision, confusion, intractable seizures as well as permanent neurological damage, probably affects more people than cancer. Trans-nonachlor and oxychlordane in serum of mothers during gestation has been linked with behaviors associated with autism in offspring at age 4-5. The Agency for Toxic Substances and Disease Registry (ATSDR) has defined a concentration of chlordane compounds of 20 ng/m^3 as the Minimal Risk Level (MRLs). ATSDR defines Minimal Risk Level as an estimate of daily human exposure to a dose of a chemical that is likely to be without an appreciable risk of adverse non-cancerous effects over a specific duration of exposure.

Remediation

Chlordane was applied under the home/building during treatment for termites and the half-life can be up to 30 years. Chlordane has a low vapor pressure and volatilizes slowly into the air of home/building above. To remove chlordane from indoor air requires either ventilation (Heat Exchange Ventilation) or activated carbon filtration. Chemical remediation of chlordane in soils was attempted by the US Army Corps of Engineers by mixing chlordane with aqueous lime and persulfate. In a phytoremediation study, Kentucky bluegrass and Perennial ryegrass were found to be minimally affected by chlordane, and both were found to take it up into their roots and shoots. Mycoremediation of chlordane in soil have found that contamination levels were reduced. The fungus Phanerochaete chrysosporium has been found to reduce concentrations of chlordane by 21% in water in 30 days and in solids in 60 days.

DIELDRIN

Dieldrin has been used in agriculture for the control of soil insects and several insect vectors of disease but this latter use has been banned in a number of countries due to environmental and human health concerns. Principle contemporary uses are restricted to controlling termites and wood borers and against textile pests. Dieldrin binds strongly to soil particles and hence is very resistant to leaching into groundwater. Volatilization is an important mechanism of loss from the soil, and because of its persistent nature and hydrophobicity, Dieldrin is known to bioconcentrate.

Action to ban Dieldrin has been taken in many countries, including Bulgaria, Ecuador, the EU, Hungary, Israel, Portugal, Singapore, Sweden, and Turkey. Its use is severely restricted in numerous countries, including Argentina, Austria, Canada, Colombia, Cyprus, India, Japan, New Zealand, Pakistan, USA and Venezuela.

Aldrin and Dieldrin containers.

Effects on Humans

In a study using human volunteers, the subjects received Dieldrin daily for 2 years. All the volunteers continued in excellent health, and clinical, physiological and laboratory findings remained essentially unchanged through the exposure period and eight-month follow up. In a study of workers from a plant involved in the manufacture of Aldrin, Dieldrin and Endrin, a statistically significant increase in liver and biliary tract cancers was observed, although the study did have some limitations, including lack of quantitative exposure information.

Effects on Plants and Animals

In laboratory studies, acute oral LD50 values in the range of 37 mg/kg body weight in rats to 330 mg/kg in hamsters have been found for Dieldrin. As with other organochlorine compounds, the liver is the major target organ in rats, with effects that included increased liver/body weight ratio, hypertrophy and histopathological changes. The no observed adverse effect level (NOAEL) in rats is 0.5 mg/kg diet, equal to 0.025 mg/kg body weight/day. When rats were fed Dieldrin in their diet over three generations, no changes in reproductive capacity were observed at any dose level tested. A NOAEL of 2 mg Dieldrin/kg diet has been set for reproduction in rats. There was no evidence for teratogenic potential in studies in rats, mice or rabbits using oral doses of up to 6 mg/kg body weight. Abnormal development and fetotoxicity were observed in hamsters and mice, however, these results are unlikely to be of significance in view of the maternal toxicity noted at the high dose levels. There is limited evidence that cyclodienes such as Dieldrin may affect immune responses. IARC has concluded that there is inadequate evidence of the carcinogenicity of Dieldrin in humans, limited evidence in experimental animals and has been classified by IARC in Group 3.

Dieldrin has low phytotoxicity. Plants are affected only by application rates much higher than suggested use rates. The acute toxicity of Dieldrin is quite variable for aquatic invertebrates, with insects being the most sensitive group (values range from 0.2-40 µg/L). It is highly toxic to most species of fish tested in the laboratory (values range from 1.1-41 µg/L). Acute toxicity of Dieldrin in frogs (96-h LC50) ranged from 8.7 µg/L for Rana catesbeiana tadpoles to 71.3µg/L for the tadpoles of Rana pipiens. Spinal deformities in embryo-larval tests were observed at concentrations as low as 1.3 µg/L for Xenopus laevis after a 10 day exposure.

The acute toxicity of Dieldrin to avian species varies widely, with acute oral LD50 values in the range of 26.6 mg/kg in pigeons to 381 mg/kg in mallard ducks. Mallard ducklings were exposed to Dieldrin in the diet for 24 days. A 24 NOAEL of 0.3µg Dieldrin/g diet, based on growth impairment was determined. Reproduction success in birds has not been consistently affected in the absence of maternal toxicity.

The acute LD50 of Dieldrin to four species of voles range from 100 to 210 mg/kg body weight, suggesting that these microtine rodents are less susceptible than laboratory rodents to Dieldrin. In another study, white tailed deer (Odocoileus virginianus) were fed diet containing Dieldrin for up to 3 years. Adult survival was not affected, and fertility and in utero mortality was comparable for all groups. Fawns from treated does were smaller at birth, experienced greater postpartum mortality and weight gain was reduced. Blesbuck (Damaliscuc dorcas phillipsi) were fed diets containing Dieldrin for 90 days. None of the animals fed 5 or 15 mg/kg diet died during the study period, but all animals at the higher dose levels died within 24 days.

The half life of Dieldrin in temperate soils is approximately 5 years. This persistence, combined with high lipid solubility, provides the necessary conditions for Dieldrin to bioconcentrate and biomagnify in organisms. Bioconcentration factors of 12,500 and 13,300 have been reported for guppies and sculpins, respectively. It is likely that Dieldrin is bioconcentrated by aquatic organisms rather than bioaccumulated. Dieldrin's chemical properties (low water solubility, high stability, and semi-volatility) favour its long range transport, and Dieldrin has been detected in Arctic air, water and organisms.

Accumulation Studies

Dieldrin residues have been detected in air, water, soil, fish, birds and mammals, including humans and human breast milk. As Aldrin is readily and rapidly converted to Dieldrin in the environment and in organisms, the levels of Dieldrin detected likely reflect the total concentrations of both compounds.

- In Egypt, the estimated dietary intake of Dieldrin by breast fed infants is of 1.22 μg/kg body weight/ day. Diet is the main source of exposure to the general public.
- In the USA, Dieldrin was the second most common pesticide detected in a survey of US pasteurized milk, detected in 172 of the 806 composite samples tested, with a maximum level of 0.003 ppm.
- In Canada, Dieldrin residues were detected in 9 of 602 (1.5%) samples of domestic animal fats and eggs, with a maximum of 0.050 mg/kg.
- In Spain, Dieldrin was detected in meat, with residues of 20 to 40 ppb were detected in the fat of 8 to 15% of pork products (meat, cured sausage, pork bologna), and in 28% fresh poultry sausage. Dieldrin residues were detected in Oriental party beans at 3.45 ppb.
- The average daily intake of Aldrin and Dieldrin in India was calculated to be 19 μg/person, exceeding the acceptable daily intake of 6.0 μg/60 kg of body weight recommended by the Joint FAO/WHO Meeting on Pesticide Residues (JMPR). Dairy products, such as milk and butter, and animal meats were the primary sources of exposure.
- Exposure through food intake has been estimated at 0.55 μg/person in Vietnam.

ENDRIN

Endrin is an organochloride with the chemical formula $C_{12}H_8Cl_6O$ that was first produced in 1950 by Shell and Velsicol Chemical Corporation. It was primarily used as an insecticide, as well as a rodenticide and piscicide. It is a colourless, odorless solid, although commercial samples are often off-white. Endrin was manufactured as an emulsifiable solution known commercially as Endrex. The compound became infamous as a persistent organic pollutant and for this reason it is banned in many countries.

In the environment endrin exists as either endrin aldehyde or endrin ketone and can be found mainly in bottom sediments of bodies of water. Exposure to endrin can occur by inhalation, ingestion of substances containing the compound, or skin contact. Upon entering the body, it can be stored in body fats and can act as a neurotoxin on the central nervous system, which can cause convulsions, seizures, or even death.

Although endrin is not currently classified as a mutagen, nor as a human carcinogen, it is still a toxic chemical in other ways with detrimental effects. Due to these toxic effects, the manufacturers cancelled all use of endrin in the United States by 1991. Food import concerns have been raised because some countries may have still been using endrin as a pesticide.

Production

The synthesis of endrin begins with the condensation of hexachlorocyclopentadiene with vinyl chloride. The product is then dehydrochlorinated. Following reaction with cyclopentadiene, isodrin is formed. Epoxide formation by adding either peracetic acid or perbenzoic acid to the isodrin is the final step in synthesizing endrin.

Endrin is a stereoisomer of dieldrin with comparable properties, though endrin degrades more easily.

Use

Endrin was formulated as emulsifiable concentrates (ECs), wettable powders (WPs), granules, field strength dusts (FSDs), and pastes. The product could then be applied by aircraft or by handheld sprayers in its various formulations.

Endrin has been used primarily as an agricultural insecticide on tobacco, apple trees, cotton, sugar cane, rice, cereal, and grains. It is effective against a variety of species, including cotton bollworms, corn borers, cut worms and grass hoppers. In addition, endrin has been employed as a rodenticide and avicide. In Malaysia, fish farms used a solution of endrin as a piscicide to rid mine pools and fish ponds of all fish prior to restocking.

A study conducted from 1981 to 1983 in the US aimed to determine endrin's effects on non-target organisms when applied as a rodenticide in orchards. Most wildlife in and around the orchard was found to have endrin exposure, with endrin toxicity accounting for more than 24% of bird deaths recorded. Endrin was eventually banned in the US on October 10, 1984.

Health Effects

- Exposure and metabolism: Exposure to endrin can occur by inhalation, ingestion of substances containing the compound, or by skin contact. In addition to inhalation and skin contact, infants can be exposed by ingesting the breast milk of an exposed woman. In utero, fetuses are exposed by way of the placenta if the mother has been exposed.

  Upon entering the body, endrin metabolizes into *anti*-12-hydroxyendrin and other metabolites, which can be expelled in the urine and feces. Both *anti*-12-hydroxyendrin and its metabolite, 12-ketoendrin, are likely responsible for the toxicity of endrin. The rapid metabolism of endrin into these metabolites makes detection of endrin itself difficult unless exposure is very high.

- Neurological effects: Symptoms of endrin poisoning include headache, dizziness, nervousness, confusion, nausea, vomiting, and convulsions. Acute endrin poisoning in humans affects primarily the central nervous system. There, it can act as a neurotoxin that blocks the activity of inhibitory neurotransmitters. In cases of acute exposure, this may result in seizures, or even death. Because endrin can be stored in body fats, acute endrin poisoning can lead to recurrent seizures when stressors induce the release of endrin back into the body, even months after the initial exposure is terminated.

  People occupationally exposed to endrin may experience abnormal EEG readings even if they exhibit none of the clinical symptoms, possibly due to injury to the brain stem. These readings show bilateral synchronous theta waves with synchronous spike-and-wave complexes. EEG readings can take up to one month to return to normal.

- Developmental effects: Though endrin exposure has not been found to adversely affect fertility in mammals, an increase in fetal mortality has been observed in mice, rats, and mallard ducks. In those animals that have survived gestation, developmental abnormalities have been observed, particularly in rodents whose mothers were exposed to endrin early in pregnancy. In hamsters, the number of cases of fused ribs, cleft palate, open eyes, webbed feet, and meningoencephaloceles have increased. Along with open eyes and cleft palate, mice have developed with fused ribs and exencephaly. Skeletal abnormalities in rodents have also been reported.

- Other effects: Higher doses of endrin have been found to cause the following in rodents: renal tubular necrosis; inflammation of the liver, fatty liver, and liver necrosis; possible kidney degradation; and a decrease in body weight and body weight gain.

  Endrin is very toxic to aquatic organisms, namely fish, aquatic invertebrates, and phytoplankton. It was found to remain in the tissues of infected fish for up to one month.

Environmental Behavior

Insecticides like dieldrin and endrin have been shown to persist for decades in the environment. A definitive detection of the residues was not possible until 1971 when mass spectrometer started being used as a detector in gas chromatography. Detection of these chemicals in the environment has been reported across the world up to 2005, even though the frequency of reported cases are low due to its relatively small-scale use and very low concentrations.

Endrin regularly enters the environment when applied to crops or when rain washes it off. It has been found in water, sediments, atmospheric air and biotic environment, even after uses have been stopped. Organochlorine pesticides strongly resist degradation, are poorly soluble in water but highly soluble in lipids, which is called lipophilic. This leads to bioaccumulation in fatty tissues of organisms, mainly those dwelling in water. A high bioconcentration factor of 1335-10,000 has been reported in fish. Endrin binds very strongly to organic matter in soil and aquatic sediments due to their high adsorption coefficient, making it less likely to leach into groundwater, even though contaminated groundwater samples have been found. In 2009, EPA released data indicating that the endrin in soil could last up to 14 years or more. The extent of endrin's persistence depends highly on local conditions. For example, high temperature (230 °C) or intense sunlight leads to more rapid breakdown of endrin into endrin ketone and endrin aldehyde, however, this breakdown is less than 5%.

Removal from the Environment

In the United States, endrin was mainly disposed in land until U.S. federal regulations were applied in 1987 on land disposal of wastes containing endrin. Primary methods of endrin disappearance from soil are volatilization and photodecomposition. Under ultraviolet light, endrin forms δ-ketoendrin and International Programme on Chemical Safety (IPCS) claims that in intense summer sun, about 50% of endrin is isomerized to δ-ketoendrin in 7 days. In anaerobic conditions microbial degradation by fungi and bacteria takes place to form the same major end product.

Hazardous Substances Data Bank (HSDB) lists reductive dechlorination and incineration for field disposal of small quantities of endrin. In reductive dechlorination, endrin's chlorine atoms were completely replaced with hydrogen atoms, which is suspected to be more environmentally acceptable. Even though endrin binds very strongly to soil, phytoremediation has been proposed by group of Japanese scientists using crops in the family Cucurbitaceae. As of 2009, exact mechanisms behind the plant uptake of endrin have not been understood. Research in uptake mechanisms and factors that influence the uptake is needed for practical application.

HEPTACHLOR

Heptachlor is an organochlorine compound that was used as an insecticide. Usually sold as a white or tan powder, heptachlor is one of the cyclodiene insecticides. In 1962, Rachel Carson's Silent Spring questioned the safety of heptachlor and other chlorinated insecticides. Due to its highly stable structure, heptachlor can persist in the environment for decades. In the United States, the Environmental Protection Agency has limited the sale of heptachlor products to the specific application of fire ant control in underground transformers. The amount that can be present in different foods is regulated.

Synthesis

Analogous to the synthesis of other cyclodienes, heptachlor is produced via the Diels-Alder reaction of hexachlorocyclopentadiene and cyclopentadiene. The resulting adduct is chlorinated followed by treatment with hydrogen chloride in nitromethane in the presence of aluminum trichloride or with iodine monochloride.

Compared to chlordane, it is about 3–5 times more active as an insecticide, but more inert chemically, being resistant to water and caustic alkalies.

Metabolism

Soil microorganisms transform heptachlor by epoxidation, hydrolysis, and reduction. When the compound was incubated with a mixed culture of organisms, chlordene (hexachlorocyclopentadine, its precursor) formed, which was further metabolized to chlordene epoxide. Other metabolites include 1-hydroxychlordene, 1-hydroxy-2,3-epoxychlordene, and heptachlor epoxide. Soil microorganisms hydrolyze heptachlor to give ketochlordene. Rats metabolize heptachlor to the epoxide 1-exo-1-hydroxyheptachlor epoxide and 1,2-dihydrooxydihydrochlordene. When heptachlor epoxide was incubated with microsomal preparations form liver of pigs and from houseflies, the products found were diol and 1-hydroxy-2,3-epoxychlordene. The metabolic scheme in rats shows two pathways with the same metabolite. The first involves following scheme: heptachlor → heptachlor epoxide → dehydrogenated derivative of 1-exo-hydroxy-2,3-exo-epoxychlordene → 1,2-dihydrooxydihydrochlordene.

The second involves: heptachlor → 1-exo-hydroxychlordene → 1-exo-hydroxy, 2,3-exo-epoxychlordene → 1,2-dihydrooxydihydrochlordene.

Environmental Impact

Heptachlor is persistent organic pollutant (POP). It has a half life of ~1.3-4.2 days (air), ~0.03-0.11 years (water), and ~0.11-0.34 years (soil). One study described its half life to be 2 years and claimed that its residues could be found in soil 14 years after its initial application. Like other POPs, heptachlor is lipophilic and poorly soluble in water (0.056 mg/L at 25 °C), thus it tends to accumulate in the body fat of humans and animals.

Heptachlor epoxide is more likely to be found in the environment than its parent compound. The epoxide also dissolves more easily in water than its parent compound and is more persistent. Heptachlor and its epoxide absorb to soil particles and evaporate.

Toxicity of Heptachlor and Related Derivatives

The range of oral rat LD_{50} values are 40 mg/kg to 162 mg/kg. Daily oral doses of heptachlor at 50 and 100 mg/kg were found to be lethal to rats after 10 days. For heptachlor epoxide, the oral LD_{50} values ranging from 46.5 to 60 mg/kg. With rat oral of LD_{50} 47mg/kg, heptachlor epoxide is more toxic. A product of hydrogenation of heptachlor, β-dihydroheptachlor, has high insecticidal activity and low mammalian toxicity, rat oral LD_{50}>5,000mg/kg.

Human Impact

Humans may be exposed to heptachlor through drinking water and foods, including breast milk. Heptachlor epoxide is derived from a pesticide that was banned in the U.S. in the 1980s. It is still found in soil and water supplies and can turn up in food. It can be passed along in breast milk.

The International Agency for Research on Cancer and the EPA have classified the compound as a possible human carcinogen. Animals exposed to heptachlor epoxide during gestation and infancy are found to have changes in nervous system and immune function. Exposure to higher doses of heptachlor in newborn animals leads to decreased body weight and death.

The U.S. EPA MCL for drinking water is 0.0004 mg/L for heptachlor and 0.0002 mg/L for heptachlor epoxide. The U.S. FDA limit on food crops is 0.01 ppm, in milk 0.1 ppm, and on edible seafoods 0.3 ppm. The Occupational Safety and Health Administration has limit of 0.5 mg/m^3 (cubic meter of workplace air) for 8-hour shifts and 40-hour work weeks.

An ATSDR report in 1993 found no studies with respect to death in humans after oral exposure to heptachlor or heptachlor epoxide.

HEXACHLOROBENZENE

Hexachlorobenzene, or perchlorobenzene, is an organochloride. It is a fungicide formerly used as a seed treatment, especially on wheat to control the fungal disease bunt. It has been banned globally under the Stockholm Convention on Persistent Organic Pollutants.

Physical and Chemical Properties

HCB is a white crystalline solid that has negligible solubility in water (0.00000002 M) but sparingly soluble in organic solvents. It is most soluble in halogenated solvents like chloroform (approx 0.03 M), less soluble in esters and hydrocarbons (approx 0.020 M), and even less soluble in short chain alcohols (0.002-0.006 M). Its vapour pressure is 1.09×10^{-5} mmHg (1.45 mPa) at 20 °C. Its flash point is 242 °C and it sublimes at 322 °C.

Safety

Hexachlorobenzene is an animal carcinogen and is considered to be a probable human carcinogen. After its introduction as a fungicide in 1945, for crop seeds, this toxic chemical was found in all food types. Hexachlorobenzene was banned from use in the United States in 1966.

This material has been classified by the International Agency for Research on Cancer (IARC) as a Group 2B carcinogen (possibly carcinogenic to humans). Animal carcinogenicity data for hexachlorobenzene show increased incidences of liver, kidney (renal tubular tumours) and thyroid cancers. Chronic oral exposure in humans has been shown to give rise to a liver disease (porphyria cutanea tarda), skin lesions with discoloration, ulceration, photosensitivity, thyroid effects, bone effects and loss of hair. Neurological changes have been reported in rodents exposed to hexachlorobenzene. Hexachlorobenzene may cause embryolethality and teratogenic effects. Human and animal studies have demonstrated that hexachlorobenzene crosses the placenta to accumulate in foetal tissues and is transferred in breast milk.

HCB is very toxic to aquatic organisms. It may cause long term adverse effects in the aquatic environment. Therefore, release into waterways should be avoided. It is persistent in the environment. Ecological investigations have found that biomagnification up the food chain does occur. Hexachlorobenzene has a half life in the soil of between 3 and 6 years. Risk of bioaccumulation in an aquatic species is high.

Toxicology

- Oral LD50 (rat): 10,000 mg/kg.

- Oral LD50 (mice): 4,000 mg/kg.
- Inhalation LC50 (rat): 3600 mg/m^3.

Material has relatively low acute toxicity but is toxic because of its persistent and cumulative nature in body tissues in rich lipid content.

In Anatolia, Turkey between 1955 and 1959, during a period when bread wheat was unavailable, 500 people were fatally poisoned and more than 4,000 people fell ill by eating bread made with HCB-treated seed that was intended for agriculture use. Most of the sick were affected with a liver condition called porphyria cutanea tarda, which disturbs the metabolism of hemoglobin and results in skin lesions. Almost all breastfeeding children under the age of two, whose mothers had eaten tainted bread, died from a condition called "pembe yara" or "pink sore", most likely from high doses of HCB in the breast milk. In one mother's breast milk the HCB level was found to be 20 parts per million in lipid, approximately 2,000 times the average levels of contamination found in breast-milk samples around the world. Follow-up studies 20 to 30 years after the poisoning found average HCB levels in breast milk were still more than seven times the average for unexposed women in that part of the world (56 specimens of human milk obtained from mothers with porphyria, average value was 0.51 ppm in HCB-exposed patients compared to 0.07 ppm in unexposed controls), and 150 times the level allowed in cow's milk.

In the same follow-up study of 252 patients (162 males and 90 females, avg. current age of 35.7 years), 20–30 years postexposure, many subjects had dermatologic, neurologic, and orthopedic symptoms and signs. The observed clinical findings include scarring of the face and hands (83.7%), hyperpigmentation (65%), hypertrichosis (44.8%), pinched faces (40.1%), painless arthritis (70.2%), small hands (66.6%), sensory shading (60.6%), myotonia (37.9%), cogwheeling (41.9%), enlarged thyroid (34.9%), and enlarged liver (4.8%). Urine and stool porphyrin levels were determined in all patients, and 17 have at least one of the porphyrins elevated. Offspring of mothers with three decades of HCB-induced porphyria appear normal.

MIREX

Mirex is an organochloride that was commercialized as an insecticide and later banned because of its impact on the environment. This white crystalline odorless solid is a derivative of cyclopentadiene. It was popularized to control fire ants but by virtue of its chemical robustness and lipophilicity it was recognized as a bioaccumulative pollutant. The spread of the red imported fire ant was encouraged by the use of Mirex, which also kills native ants that are highly competitive with the fire ants. The United States Environmental Protection Agency prohibited its use in 1976. It is prohibited by the Stockholm Convention on Persistent Organic Pollutants.

Production and Applications

Mirex was first synthesized in 1946, but was not used in pesticide formulations until 1955. Mirex was produced by the dimerization of hexachlorocyclopentadiene in the presence of aluminium chloride.

Mirex is a stomach insecticide, meaning that it must be ingested by the organism in order to poison it. The insecticidal use was focused on Southeastern United States to control the imported fire ants Solenopsis saevissima richteri and Solenopsis invicta. Approximately 250,000 kg of mirex was applied to fields between 1962-75 (US NRC, 1978). Most of the mirex was in the form of "4X mirex bait," which consists of 0.3% mirex in 14.7% soybean oil mixed with 85% corncob grits. Application of the 4X bait was designed to give a coverage of 4.2 g mirex/ha and was delivered by aircraft, helicopter or tractor. 1x and 2x bait were also used. Use of mirex as a pesticide was banned in 1978. The Stockholm Convention banned production and use of several persistent organic pollutant, and Mirex is one of the "dirty dozen".

Degradation

Characteristic of chlorocarbons, mirex does not burn easily; combustion products are expected to include carbon dioxide, carbon monoxide, hydrogen chloride, chlorine, phosgene, and other organochlorine species. Slow oxidation produces chlordecone ("Kepone"), a related insecticide that is also banned in most of the western world, but more readily degraded. Sunlight degrades mirex primarily to photomirex (8-monohydromirex) and later partly to 2,8-dihydromirex.

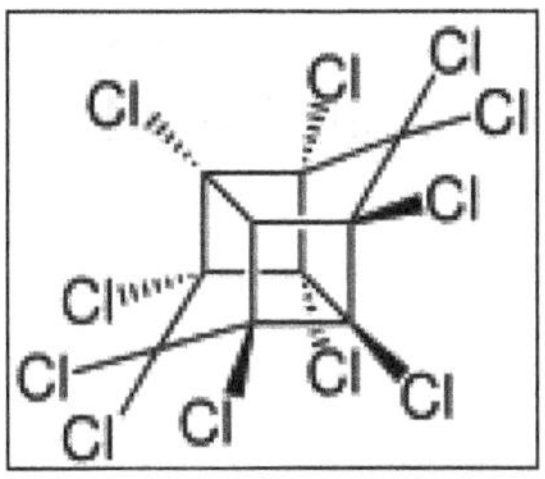

Photomirex.

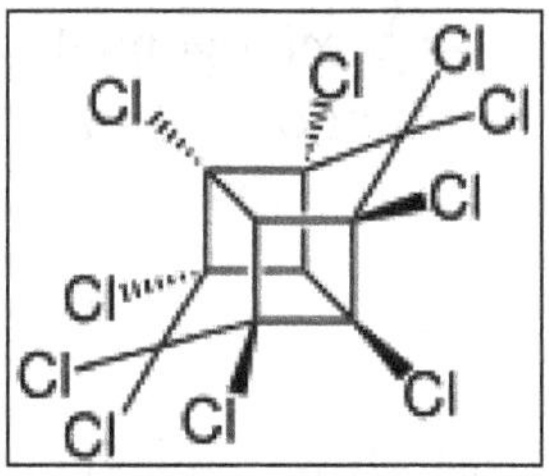

2,8-dihydromirex.

Mirex is highly resistant to microbiological degradation. It only slowly dechlorinates to a monohydro derivative by anaerobic microbial action in sewage sludge and by enteric bacteria. Degradation by soil microorganisms has not been described.

Bioaccumulation and Biomagnification

Mirex is highly cumulative and amount depends upon the concentration and duration of exposure. There is evidence of accumulation of mirex in aquatic and terrestrial food chains to harmful levels. After 6 applications of mirex bait at 1.4 kg/ha, high mirex levels were found in some species; turtle fat contained 24.8 mg mirex kg, kingfishers, 1.9 mg/kg, coyote fat, 6 mg/kg, opossum fat, 9.5 mg/kg, and racoon fat, 73.9 mg/kg. In a model ecosystem with a terrestrial-aquatic interface, sorgum seedlings were treated with mirex at 1.1 kg/ha. Caterpillars fed on these seedlings and their faeces contaminated the water which contained algae, snails, Daphnia, mosquito larvae, and fish. After 33 days, the ecological magnification value was 219 for fish and 1165 for snails.

Although general environmental levels are low, it is widespread in the biotic and abiotic environment. Being lipophilic, Mirex is strongly adsorbed on sediments.

Safety

Mirex is only moderately toxic in single-dose animal studies (oral LD_{50} values range from 365–3000 mg/kg body weight). It can enter the body via inhalation, ingestion, and via the skin. The most sensitive effects of repeated exposure in animals are principally associated with the liver, and these effects have been observed with doses as low as 1.0 mg/kg diet (0.05 mg/kg body weight per day), the lowest dose tested. At higher dose levels, it is fetotoxic (25 mg/kg in diet) and teratogenic (6.0 mg/kg per day). Mirex was not generally active in short-term tests for genetic activity. There is sufficient evidence of its carcinogenicity in mice and rats. Delayed onset of toxic effects and mortality is typical of mirex poisoning. Mirex is toxic for a range of aquatic organisms, with crustacea being particularly sensitive.

Mirex induces pervasive chronic physiological and biochemical disorders in various vertebrates. No acceptable daily intake (ADI) for Mirex has been advised by FAO/WHO. IARC (1979) evaluated mirex's carcinogenic hazard and concluded that "there is sufficient evidence for its carcinogenicity to mice and rats. In the absence of adequate data in humans, based on above result it can be said, that it has carcinogenic risk to humans". Data on human health effects do not exist.

Health Effects

Per a 1995 ATSDR report Mirex caused fatty changes in the livers, hyperexcitability and convulsion, and inhibition of reproduction in animals. It is a potent endocrine disruptor, interfering with estrogen-mediated functions such as ovulation, pregnancy, and endometrial growth. It also induced liver cancer by interaction with estrogen in female rodents.

TOXAPHENE

Toxaphene was an insecticide used primarily for cotton in the southern United States during the late 1960s and 1970s. Toxaphene is a mixture of over 670 different chemicals and is produced by reacting chlorine gas with camphene. It can be most commonly found as a yellow to amber waxy solid.

Toxaphene was banned in the United States in 1990 and was banned globally by the 2001 Stockholm Convention on Persistent Organic Pollutants. It is a very persistent chemical that can remain in the environment for 1–14 years without degrading, particularly in the soil.

Testing performed on animals, mostly rats and mice, has demonstrated that toxaphene is harmful to animals. Exposure to toxaphene has proven to stimulate the central nervous system, as well as induce morphological changes in the thyroid, liver, and kidneys.

Toxaphene has been shown to cause adverse health effects in humans. The main sources of exposure are through food, drinking water, breathing contaminated air, and direct contact with contaminated soil. Exposure to high levels of toxaphene can cause damage to the lungs, nervous system, liver, kidneys, and in extreme cases, may even cause death. It is thought to be a potential carcinogen in humans, though this has not yet been proven.

Composition

Toxaphene is a synthetic organic mixture composed of over 670 chemicals, formed by the chlorination of camphene ($C_{10}H_{16}$) to an overall chlorine content of 67–69% by weight. The bulk of the compounds (mostly chlorobornanes, chlorocamphenes, and other bicyclic chloroorganic compounds) found in toxaphene have chemical formulas ranging from $C_{10}H_{11}Cl_5$ to $C_{10}H_6Cl_{12}$, with a mean formula of $C_{10}H_{10}Cl_8$. The formula weights of these compounds range from 308 to 551 grams/mole; the theoretical mean formula has a value of 414 grams/mole. Toxaphene is usually seen as a yellow to amber waxy solid with a piney odor. It is highly insoluble in water but freely soluble in aromatic hydrocarbons and readily soluble in aliphatic organic solvents. It is stable at room temperature and pressure. It is volatile enough to be transported for long distances through the atmosphere.

Applications

Toxaphene was primarily used as a pesticide for cotton in the southern United States during the late 1960s and 1970s. It was also used on corn, small grains, vegetables, and soybeans to control ectoparasites such as lice, flies, ticks, mange, and scam mites on livestock. In some cases it was used to kill undesirable fish species in lakes and streams. The breakdown of usage can be summarized: 85% on cotton, 7% to control insect pests on livestock and poultry, 5% on other field crops, 3% on soybeans, and less than 1% on sorghum.

The first recorded usage of toxaphene was in 1966 in the United States, and by the early to mid 1970s, toxaphene was the United States' most heavily used pesticide. Over 34 million pounds of toxaphene were used annually from 1966 to 1976. As a result of Environmental Protection Agency restrictions, annual toxaphene usage fell to 6.6 million pounds in 1982. In 1990, the EPA banned all usage of toxaphene in the United States. Toxaphene is still used in countries outside the United States but much of this usage has been undocumented. Between 1970 and 1995, global usage of toxaphene was estimated to be 670 million kilograms (1.5 billion pounds).

Production

Toxaphene was first produced in the United States in 1947 although it was not heavily used until 1966. By 1975, toxaphene production reached its peak at 59.4 million pounds annually. Production decreased more than 90% from this value by 1982 due to Environmental Protection Agency restrictions. Overall, an estimated 234,000 metric tons (over 500 million pounds) have been produced in the United States. Between 25% and 35% of the toxaphene produced in the United States has been exported. There are currently 11 toxaphene suppliers worldwide.

Environmental Effects

When released into the environment, toxaphene can be quite persistent and exists in the air, soil, and water. In water, it can evaporate easily and is fairly insoluble. Its solubility is 3 mg/L of water at 22 degrees Celsius. Toxaphene breaks down very slowly and has a half-life of up to 12 years in the soil. It is most commonly found in air, soil, and sediment found at the bottom of lakes or streams. It can also be present in many parts of the world where it was never used because toxaphene is able to evaporate and travel long distances through air currents. Toxaphene can eventually be degraded, through dechlorination, in the air using sunlight to break it down. The degradation of toxaphene usually occurs under aerobic conditions. The levels of toxaphene have decreased since its ban, however, due to its persistence can still be found in the environment today.

Exposure

The three main paths of exposure to toxaphene are ingestion, inhalation, and absorption. For humans, the main source of toxaphene exposure is through ingested seafood. When toxaphene enters the body, it usually accumulates in fatty tissues. It is broken down through dechlorination and oxidation in the liver, and the byproducts are eliminated through feces.

People that live near an area that has high toxaphene contamination are at high risk to toxaphene exposure through inhalation of contaminated air or direct skin contact with contaminated soil or water. Eating large quantities of fish on a daily basis also increases susceptibility to toxaphene exposure. Finally, exposure is rare, yet possible through drinking water when contaminated by toxaphene runoff from the soil. However, toxaphene has been rarely seen at high levels in drinking water due to toxaphene's high levels of insolubility in water.

Shellfish, algae, fish and marine mammals have all been shown to exhibit high levels of toxaphene. People in the Canadian Arctic, where a traditional diet consists of fish and marine animals, have been shown to consume ten times the accepted daily intake of toxaphene. Also, blubber from beluga whales in the Arctic were found to have unhealthy and toxic levels of toxaphene.

Health Effects

In Humans

When inhaled or ingested, sufficient quantities of toxaphene can damage the lungs, nervous system, and kidneys, and may cause death. The major health effects of toxaphene involve central nervous system stimulation leading to convulsive seizures. The dose necessary to induce nonfatal convulsions in humans is about 10 milligrams per kilogram body weight per day. Several deaths linked to toxaphene have been documented in which an unknown quantity of toxaphene was ingested intentionally or accidentally from food contamination. The deaths are attributed to respiratory failure resulting from seizures. Chronic inhalation exposure in humans results in reversible respiratory toxicity.

A study conducted between 1954 and 1972 of male agricultural workers and agronomists exposed to toxaphene and other pesticides showed that there are higher proportions of bronchial carcinoma in the test group than in the unexposed general population. However, toxaphene may not have been the main pesticide responsible for tumor production. Tests on lab animals show that toxaphene causes liver and kidney cancer, so the EPA has classified it as a Group B2 carcinogen, meaning it is a probable human carcinogen. The International Agency for Research on Cancer has classified it as a Group 2B carcinogen.

Toxaphene can be detected in blood, urine, breast milk, and body tissues if a person has been exposed to high levels, but it is removed from the body quickly, so detection has to occur within several days of exposure.

It is not known whether toxaphene can affect reproduction in humans.

In Animals

Toxaphene was used to treat mange in cattle in California in the 1970s and there were reports of cattle deaths following the toxaphene treatment.

Chronic oral exposure in animals affects the liver, the kidney, the spleen, the adrenal and thyroid glands, the central nervous system, and the immune system. Toxaphene stimulates the central nervous system by antagonizing neurons leading to hyperpolarization of neurons and increased neuronal activity.

POLYCHLORINATED DIBENZODIOXINS

Polychlorinated dibenzodioxins (PCDDs), or simply dioxins, are a group of polyhalogenated organic compounds that are significant environmental pollutants.

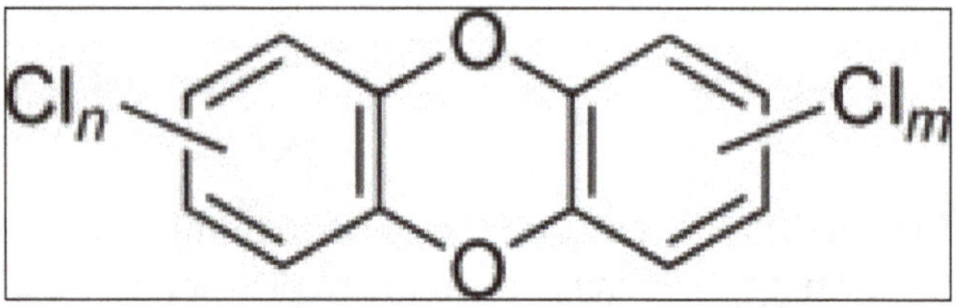

General structure of PCDDs where *n* and *m* can range from 0 to 4.

They are commonly but inaccurately referred to as dioxins for simplicity, because every PCDD molecule contains a dibenzo-1,4-dioxin skeletal structure, with 1,4-dioxin as the central ring. Members of the PCDD family bioaccumulate in humans and wildlife because of their lipophilic properties, and may cause developmental disturbances and cancer.

Dioxins occur as by-products in the manufacture of some organochlorides, in the incineration of chlorine-containing substances such as polyvinyl chloride (PVC), in the chlorine bleaching of paper, and from natural sources such as volcanoes and forest fires. There have been many incidents of dioxin pollution resulting from industrial emissions and accidents; the earliest such incidents were in the mid 19th century during the Industrial Revolution.

Chemical Structure of Dibenzo-1,4-dioxins

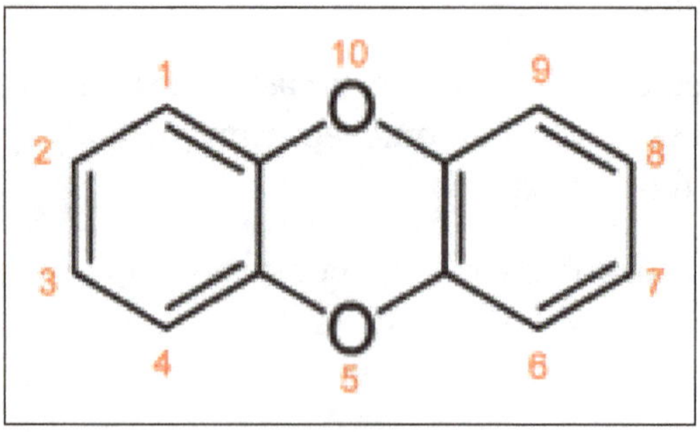

The skeletal formula and substituent numbering scheme of the parent compound dibenzo-1,4-dioxin.

The structure of dibenzo-1,4-dioxin consists of two benzene rings joined by two oxygen bridges. This makes the compound an aromatic diether. The name dioxin formally refers to the central dioxygenated ring, which is stabilized by the two flanking benzene rings.

In PCDDs, chlorine atoms are attached to this structure at any of 8 different places on the molecule, at positions 1–4 and 6–9. There are 75 different PCDD congeners (that is, related dioxin compounds).

The toxicity of PCDDs depends on the number and positions of the chlorine atoms. Congeners that have chlorine in the 2, 3, 7, and 8 positions have been found to be significantly toxic. In fact, 7 congeners have chlorine atoms in the relevant positions which were considered toxic by the World Health Organization toxic equivalent (WHO-TEQ) scheme.

Sources of Dioxins

The United States Environmental Protection Agency inventory of sources of dioxin-like compounds is possibly the most comprehensive review of the sources and releases of dioxins, but other countries now have substantial research as well.

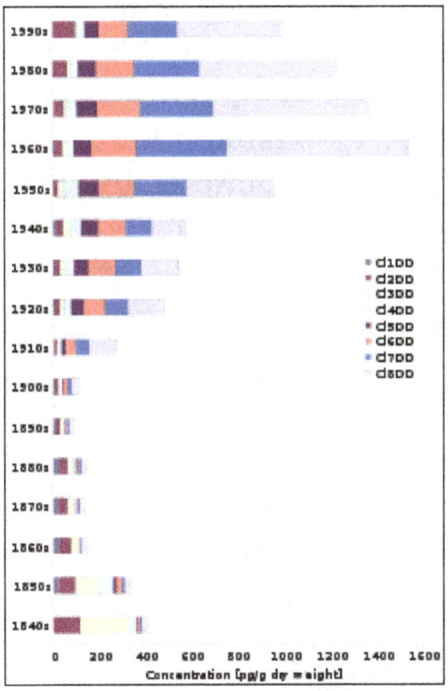

Concentration profile of PCDD in a dated sediment core.

Occupational exposure is an issue for some in the chemical industries, historically for those making chlorophenols or chlorophenoxy acid herbicides or in the application of chemicals, notably herbicides. In many developed nations there are now emissions regulations which have dramatically decreased the emissions and thus alleviated some concerns, although the lack of continuous sampling of dioxin emissions causes concern about the understatement of emissions. In Belgium, through the introduction of a process called AMESA, continuous sampling showed that periodic sampling understated emissions by a factor of 30 to 50 times. Few facilities have continuous sampling.

Dioxins are produced in small concentrations when organic material is burned in the presence of chlorine, whether the chlorine is present as chloride ions or as organochlorine compounds, so they are widely produced in many contexts. According to the most recent US EPA data, the major sources of dioxins are broadly in the following types:

- Combustion sources, e.g. municipal waste or medical waste incinerators and private backyard barrel burning.
- Metal smelting.
- Refining and process sources.
- Chemical manufacturing sources.
- Natural sources.
- Environmental reservoirs.

When first carried out in 1987, the original US EPA inventory of dioxin sources revealed that incineration represented more than 80% of known dioxin sources. As a result, US EPA implemented new emissions requirements. These regulations succeeded in reducing dioxin stack emissions from incinerators. Incineration of municipal solid waste, medical waste, sewage sludge, and hazardous waste together now produce less than 3% of all dioxin emissions. Since 1987, however, backyard barrel burning has showed almost no decrease, and is now the largest source of dioxin emissions, producing about one third of the total output.

In incineration, dioxins can also reform or form de novo in the atmosphere above the stack as the exhaust gases cool through a temperature window of 600 to 200 °C. The most common method of reducing the quantity of dioxins reforming or forming de novo is through rapid (30 millisecond) quenching of the exhaust gases through that 400 °C window. Incinerator emissions of dioxins have been reduced by over 90% as a result of new emissions control requirements. Incineration in developed countries is now a very minor contributor to dioxin emissions.

Dioxins are also generated in reactions that do not involve burning — such as chlorine bleaching fibers for paper or textiles, and in the manufacture of chlorinated phenols, particularly when reaction temperature is not well controlled. Compounds involved include the wood preservative pentachlorophenol, and also herbicides such as 2,4-dichlorophenoxyacetic acid (or 2,4-D) and 2,4,5-trichlorophenoxyacetic acid (2,4,5-T). Higher levels of chlorination require higher reaction temperatures and greater dioxin production. Dioxins may also be formed during the photochemical breakdown of the common antimicrobial compound triclosan.

Sources of Human Intake

Tolerable daily, monthly or annual intakes have been set by the World Health Organization and a number of governments. Dioxins enter the general population almost

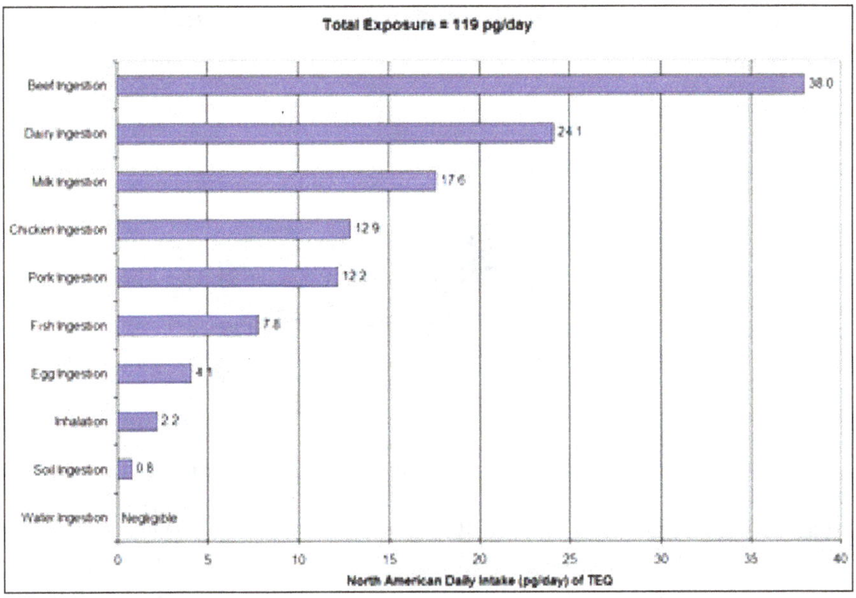

A chart illustrating how much dioxin the average American consumes per day. (pg = picogram, or one trillionth of a gram, or 10^{-12} g).

Children are passed substantial body burdens by their mothers, and breastfeeding increases the child's body burden. Dioxin exposure can also occur from contact with Pentachlorophenol (Penta) treated lumber as Pentachlorophenol often contains dioxins as a contaminant. Children's daily intakes during breast feeding are often many times above the intakes of adults based on body weight. This is why the WHO consultation group assessed the tolerable intake so as to prevent a woman from accumulating harmful body burdens before her first pregnancy. Breast fed children usually still have higher dioxin body burdens than non breast fed children. The WHO still recommends breast feeding for its other benefits. In many countries dioxins in breast milk have decreased by even 90% during the two last decades.

Dioxins are present in cigarette smoke. Dioxin in cigarette smoke was noted as "understudied" by the US EPA in its "Re-Evaluating Dioxin". In that same document, the US EPA acknowledged that dioxin in cigarettes is "anthropogenic" (man-made, "not likely in nature").

Metabolism

Dioxins are absorbed primarily through dietary intake of fat, as this is where they accumulate in animals and humans. In humans, the highly chlorinated dioxins are stored in fatty tissues and are neither readily metabolized nor excreted. The estimated elimination half-life for highly chlorinated dioxins (4–8 chlorine atoms) in humans ranges from 4.9 to 13.1 years.

The persistence of a particular dioxin congener in an animal is thought to be a consequence of its structure. Dioxins with no lateral (2, 3, 7, and 8) chlorines, which thus contain hydrogen atoms on adjacent pairs of carbons, can more readily be oxidized by cytochromes P450. The oxidized dioxins can then be more readily excreted rather than stored for a long time.

Toxicity

Space-filling model of 2,3,7,8-tetrachlorodibenzodioxin.

2,3,7,8-Tetrachlorodibenzodioxin (TCDD) is considered the most toxic of the congeners (for the mechanism of action, see 2,3,7,8-Tetrachlorodibenzodioxin and Aryl hydrocarbon receptor). Other dioxin congeners including PCDFs and PCBs with dioxin-like toxicity, are given a toxicity rating from 0 to 1, where TCDD = 1. This toxicity rating is called the Toxic Equivalence Factor concept, or TEF. TEFs are consensus values and, because of the strong species dependence for toxicity, are listed separately for mammals, fish, and birds. TEFs for mammalian species are generally applicable to human risk calculations. The TEFs have been developed from detailed assessment of literature data to facilitate both risk assessment and regulatory control. Many other compounds may also have dioxin-like properties, particularly non-ortho PCBs, one of which has a TEF as high as 0.1.

The total dioxin toxic equivalence (TEQ) value expresses the toxicity as if the mixture were pure TCDD. The TEQ approach and current TEFs have been adopted internationally as the most appropriate way to estimate the potential health risks of mixture of dioxins. Recent data suggest that this type of simple scaling factor may not be the most appropriate treatment for complex mixtures of dioxins; both transfer from the source and absorption and elimination vary among different congeners, and the TEF value is not able to accurately reflect this.

Dioxins and other persistent organic pollutants (POPs) are subject to the Stockholm Convention. The treaty obliges signatories to take measures to eliminate where possible, and minimize where not possible to eliminate, all sources of dioxin.

Health Effects in Humans

Dioxins build up primarily in fatty tissues over time (bioaccumulation), so even small exposures may eventually reach dangerous levels. In 1994, the US EPA reported that dioxins are a probable carcinogen, but noted that non-cancer effects (reproduction and sexual development, immune system) may pose a greater threat to human health.

TCDD, the most toxic of the dibenzodioxins, is classified as a Group 1 carcinogen by the International Agency for Research on Cancer (IARC). TCDD has a half-life of approximately 8 years in humans, although at high concentrations, the elimination rate is enhanced by metabolism. The health effects of dioxins are mediated by their action on a cellular receptor, the aryl hydrocarbon receptor (AhR).

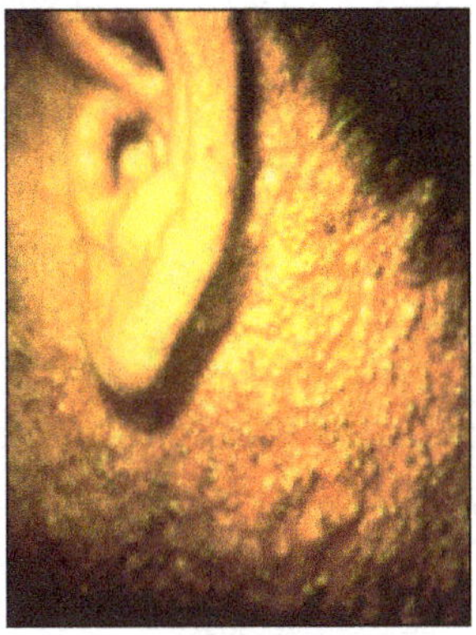

Chloracne on the ear and neck of a herbicide production worker.

Exposure to high levels of dioxins in humans causes a severe form of persistent acne, known as chloracne. High occupational or accidental levels of exposures to dioxins have been shown by epidemiological studies to lead to an increased risk of tumors at all sites. Other effects in humans (at high dose levels) may include:

- Developmental abnormalities in the enamel of children's teeth.
- Central and peripheral nervous system pathology.
- Thyroid disorders.
- Damage to the immune systems.
- Endometriosis.
- Diabetes.

Recent studies have shown that high exposure to dioxins changes the ratio of male to female births among a population such that more females are born than males.

Dioxins accumulate in food chains in a fashion similar to other chlorinated compounds (bioaccumulation). This means that even small concentrations in contaminated water can be concentrated up a food chain to dangerous levels because of the long biological half life and low water solubility of dioxins.

Toxic Effects in Animals

While it has been difficult to establish specific health effects in humans due to the lack of controlled dose experiments, studies in animals have shown that dioxin causes a wide variety of toxic effects. In particular, TCDD has been shown to be teratogenic, mutagenic, carcinogenic, immunotoxic, and hepatotoxic. Furthermore, alterations in multiple endocrine and growth factor systems have been reported. The most sensitive effects, observed in multiple species, appear to be developmental, including effects on the developing immune, nervous, and reproductive systems. The most sensitive effects are caused at body burdens relatively close to those reported in humans.

Among the animals for which TCDD toxicity has been studied, there is strong evidence for the following effects:

- Birth defects (teratogenicity): In rodents, including rats, mice, hamsters and guinea pigs, birds, and fish.
- Cancer (including neoplasms in the mammalian lung, oral/nasal cavities, thyroid and adrenal glands, and liver, squamous cell carcinoma, and various animal hepatocarcinomas): In rodents and fish.
- Hepatotoxicity (liver toxicity): In rodents, chickens, and fish.
- Endocrine disruption: In rodents and fish.
- Immunosuppression: In rodents and fish.

The LD50 of dioxin also varies wildly between species with the most notable disparity being between the ostensibly similar species of hamster and guinea pig. The oral LD50 for guinea pigs is as low as 0.5 to 2 μg/Kg body weight, whereas the oral LD50 for hamsters can be as high as 1 to 5 mg/Kg body weight, a difference of as much as thousandfold or more, and even among rat strains there may be thousandfold differences.

Agent Orange

U.S. Army Huey helicopter spraying Agent Orange over Vietnamese agricultural land.

Agent Orange was the code name for one of the herbicides and defoliants the U.S. military used as part of its herbicidal warfare program, Operation Ranch Hand, during the Vietnam War from 1961 to 1971. It was a mixture of 2,4,5-T and 2,4-D. The 2,4,5-T used was contaminated with 2,3,7,8-tetrachlorodibenzodioxin (TCDD), an extremely toxic dioxin compound.

During the Vietnam war, between 1962 and 1971, the United States military sprayed 20,000,000 U.S. gallons (76,000,000 L) of chemical herbicides and defoliants in Vietnam, eastern Laos and parts of Cambodia, as part of Operation Ranch Hand.

By 1971, 12% of the total area of South Vietnam had been sprayed with defoliating chemicals, which were often applied at rates that were 13 times as high as the legal USDA limit. In South Vietnam alone, an estimated 10 million hectares of agricultural land were ultimately destroyed. In some areas, TCDD concentrations in soil and water were hundreds of times greater than the levels considered safe by the U.S. Environmental Protection Agency.

According to Vietnamese Ministry of Foreign Affairs, 4.8 million Vietnamese people were exposed to Agent Orange, resulting in 400,000 people being killed or maimed, and 500,000 children born with birth defects. The Red Cross of Vietnam estimates that up to 1 million people are disabled or have health problems due to Agent Orange contamination. The United States government has challenged these figures as being unreliable and unrealistically high.

Dioxin Exposure Incidents

- In 1949, in a Monsanto herbicide production plant for 2,4,5-T in Nitro, West Virginia, 240 people were affected when a relief valve opened.
- In 1963, a dioxin cloud escaped after an explosion in a Philips-Duphar plant (now Solvay Group) near Amsterdam. The plant was so polluted with dioxin after the accident that it had to be dismantled, embedded in concrete, and dumped into the ocean.

Spolana Neratovice chloralkali plant, air view.

- Between 1965 and 1968 production of 2,4,5-trichlorophenol in Spolana Neratovice plant in Czechoslovakia seriously poisoned about 60 workers with dioxins; after 3 years of investigation of the health problems of workers, Spolana stopped manufacture of 2,4,5-T (most of which was supplied to the US military in Vietnam). Several buildings of the Spolana chemical plant were heavily contaminated by dioxins. Unknown amounts of dioxins were flushed into the Elbe and Mulde rivers during the 2002 European flood, contaminating soil. Analysis of eggs and ducks found dioxin levels 15 times higher than the EU limit and a high concentration of dioxin-like PCBs in the village of Libiš. In 2004, the state health authority published a study which analysed the level of toxic substances in human blood near Spolana. According to the study, dioxin levels in Neratovice, Libiš and Tišice were about twice the level of the control group in Benešov. The quantity of dioxins near Spolana is significantly higher than the background levels in other countries. According to the US EPA, even a background level can pose a risk of cancer from 1:10000 to 1:1000, about 100 times higher than normal. The consumption of local fish, eggs, poultry, and some produce was prohibited because of post-flood contamination.

- Also during 1965 through 1968, Dr. Albert M. Kligman was contracted by the Dow Chemical Company to perform threshold tests for TCDD on inmates at Holmesburg Prison in Philadelphia after Dow studies revealed adverse effects on workers at Dow's Midland, Michigan, plant were likely due to TCDD. A subsequent test by Dow in rabbit ear models when exposed to 4–8μg usually caused a severe response. The human studies carried out in Holmesburg failed to follow Dow's original protocol and lacked proper informed consent by the participants. As a result of poor study design and subsequent destruction of records, the tests were virtually worthless even though ten inmates were exposed to 7,500μg of TCDD.
- In 1976, large amounts of dioxins were released in an industrial accident at Seveso, Italy, although no immediate human fatalities or birth defects occurred.
- In 1978, dioxins were some of the contaminants that forced the evacuation of the Love Canal neighborhood of Niagara Falls, New York.
- From 1982 through to 1985, Times Beach, Missouri, was bought out and evacuated under order of the United States Environmental Protection Agency due to high levels of dioxins in the soil caused by applications of contaminated oil meant to control dust on the town's dirt roads. The town eventually disincorporated.
- A chemical plant Khimprom in Ufa, Russia released phenol into the water tributaries in the Spring of 1990. An investigation revealed previously classified disposal of dioxin in manufacturing 2,4,5-Trichlorophenoxyacetic_acid. The Spring 1990 accident affected 670,000 people. Dioxin was found in tap water. It was assumed that it resulted from chlorophenol produced by a reaction with chlorine in water purification.

- In December 1991, an electrical explosion caused dioxins (created from the oxidation of polychlorinated biphenyl) to spread through four residence halls and two other buildings on the college campus of SUNY New Paltz.

- In May 1999, there was a dioxin crisis in Belgium: quantities of polychlorinated biphenyls with dioxin-like toxicity had entered the food chain through contaminated animal feed. 7,000,000 chickens and 60,000 pigs had to be slaughtered. This scandal was followed by a landslide change in government in the elections one month later.

- Explosions resulting from the terrorist attacks on the US on September 11, 2001, released massive amounts of dust into the air. The air was measured for dioxins from September 23, 2001, to November 21, 2001, and reported to be "likely the highest ambient concentration that have ever been reported [in history]." The United States Environmental Protection Agency report dated October 2002 and released in December 2002 titled "Exposure and Human Health Evaluation of Airborne Pollution from the World Trade Center Disaster" authored by the EPA Office of Research and Development in Washington states that dioxin levels recorded at a monitoring station on Park Row near City Hall Park in New York between October 12 and 29, 2001, averaged 5.6 parts per trillion, or nearly six times the highest dioxin level ever recorded in the U.S. Dioxin levels in the rubble of the World Trade Centers were much higher with concentrations ranging from 10 to 170 parts per trillion. The report did no measuring of the toxicity of indoor air.

- In a 2001 case study, physicians reported clinical changes in a 30-year-old woman who had been exposed to a massive dosage (144,000 pg/g blood fat) of dioxin equal to 16,000 times the normal body level; the highest dose of dioxin ever recorded in a human. She suffered from chloracne, nausea, vomiting, epigastric pain, loss of appetite, leukocytosis, anemia, amenorrhoea and thrombocytopenia. However, other notable laboratory tests, such as immune function tests, were relatively normal. The same study also covered a second subject who had received a dosage equivalent to 2,900 times the normal level, who apparently suffered no notable negative effects other than chloracne. These patients were provided with olestra to accelerate dioxin elimination.

- In 2004, in a notable individual case of dioxin poisoning, Ukrainian politician Viktor Yushchenko was exposed to the second-largest measured dose of dioxins, according to the reports of the physicians responsible for diagnosing him. This is the first known case of a single high dose of TCDD dioxin poisoning, and was diagnosed only after a toxicologist recognized the symptoms of chloracne while viewing television news coverage of his condition.

- In the early 2000s, residents of the city of New Plymouth, New Zealand, reported many illnesses of people living around and working at the Dow Chemical plant. This plant ceased production of 2,4,5-T in 1987.

- DuPont has been sued by 1,995 people who claim dioxin emissions from DuPont's plant in DeLisle, Mississippi, caused their cancers, illnesses or loved ones' deaths; of these only 850 were pending as of June 2008. In August 2005, Glen Strong, an oyster fisherman with the rare blood cancer multiple myeloma, was awarded $14 million from DuPont, but the ruling was overturned June 5, 2008, by a Mississippi jury who found DuPont's plant had no connection to Mr. Strong's disease. In another case, parents claimed dioxin from pollution caused the death of their 8-year-old daughter; the trial took place in the summer of 2007, and a jury wholly rejected the family's claims, as no scientific connection could be proven between DuPont and the family's tragic loss. DuPont's DeLisle plant is one of three titanium dioxide facilities (including Edgemoor, Delaware, and New Johnsonville, Tennessee) that are the largest producers of dioxin in the country, according to the US EPA's Toxic Release Inventory. DuPont maintains its operations are safe and environmentally responsible.

- In 2007, thousands of tonnes of foul-smelling refuse were piled up in Naples, Italy and its surrounding villages, defacing entire neighbourhoods. Authorities discovered that polychlorinated dibenzodioxins levels in buffalo milk used by 29 mozzarella makers exceeded permitted limits; after further investigation they impounded milk from 66 farms. Authorities suspected the source of the contamination was from waste illegally disposed of on land grazed by buffalo. Prosecutors in Naples placed 109 people under investigation on suspicion of fraud and food poisoning. Sales of Mozzarella cheese fell by 50% in Italy.

- In December 2008 in Ireland dioxin levels in pork were disclosed to have been between 80 and 200 times the legal limit. All Irish pork products were withdrawn from sale both nationally and internationally. In this case the dioxin toxicity was found to be mostly due to dioxin-like polychlorinated dibenzofurans and polychlorinated biphenyls, and the contribution from actual polychlorinated dibenzodioxins was relatively low. It is thought that the incident resulted from the contamination of fuel oil used in a drying burner at a single feed processor, with PCBs. The resulting combustion produced a highly toxic mixture of PCBs, dioxins and furans, which was included in the feed produced and subsequently fed to a large number of pigs.

- According to data in 2009, in 2005 the production of dioxin by the steel industry ILVA in Taranto (Italy) accounted for 90.3 per cent of the overall Italian emissions, and 8.8 per cent of the European emissions.

- German dioxin incident: In January 2011 about 4700 German farms were banned from making deliveries after self-checking of an animal feed producer had showed levels of dioxin above maximum levels. This incident appeared to involve PCDDs and not PCBs. Dioxins were found in animal feed and eggs in many farms. The maximum values were exceeded twofold in feed and maximally fourfold in some individual eggs. Thus the incident was minor as compared with the Belgian crisis in 1999, and delivery bans were rapidly cleared.

Dioxin Testing

The analyses used to determine these compounds' relative toxicity share common elements that differ from methods used for more traditional analytical determinations. The preferred methods for dioxins and related analyses use high resolution gas chromatography/mass spectrometry (HRGC/HRMS). Concentrations are determined by measuring the ratio of the analyte to the appropriate isotopically labeled internal standard.

Also novel bio-assays like DR CALUX are nowadays used in identification of dioxins and dioxin-like compounds. The advantage in respect to HRGC/HRMS is that it is able to scan many samples at lower costs. Also it is able to detect all compounds that interact with the Ah-receptor which is responsible for carcinogenic effects.

POLYCHLORINATED BIPHENYLS

Polychlorinated biphenyls (PCBs) are mixtures of chlorinated hydrocarbons that have been used extensively since 1930 in a variety of industrial uses, including as dielectrics in transformers and large capacitors, as heat exchange fluids, as paint additives, in carbonless copy paper and in plastics.

Transformer oil storage drums.

The value of PCBs for industrial applications is related to their chemical inertness, resistance to heat, non-flammability, low vapour pressure, and high dielectric constant. There are 209 possible PCBs, from three monochlorinated isomers to the fully chlorinated decachlorobiphenyl isomer. Generally, the water solubility and vapour pressure decrease as the degree of substitution increases, and the lipid solubility increases with increasing chlorine substitution. PCBs in the environment may be expected to associate with the organic components of soils, sediments, and biological tissues, or with dissolved organic carbon in aquatic systems, rather than being in solution in water.

PCBs volatilize from water surfaces in spite of their low vapour pressure, and partly as a result of their hydrophobicity; atmospheric transport may therefore be a significant pathway for the distribution of PCBs in the environment.

The toxicology of PCBs is affected by the number and position of the chlorine atoms, as substitution in the ortho position hinders the rotation of the rings. PCBs without ortho substitution are generally referred to as coplanar and all others as noncoplanar. Coplanar PCBs, like dioxins and furans, bind to the AL-receptor and may exert, thus, dioxin-like effects, in addition to AL-receptor independent effects which they share with non-coplanar PCBs (e.g. tumor promoter).

Human Exposure

Association between elevated exposure to PCB mixtures and alterations in liver enzymes, hepatomegaly, and dermatological effects such as rashes and acne has been reported. Adverse effects are predominantly associated with higher blood concentrations.

Contamination of rice oil by PCBs in Japan and Taiwan has resulted in the exposure of a large number of people to PCBs and their contaminants PCDFs. Signs and symptoms of exposure from these incidents include enlargement and hyper secretion of the Meibomian glands of the eyes, swelling of the eyelids, and pigmentation of the nails and mucous membranes, occasionally associated with fatigue, nausea and vomiting. This was followed by hyperkeratosis and darkening of the skin with follicular enlargement and acneform eruptions, often with a secondary staphylococcal infection.

Children born up to 7 years after maternal exposure in the Taiwan incident had hyperpigmentation, deformed nails and natal teeth, intrauterine growth delay, poorer cognitive development up to 7 years of age, behavioural problems and higher activity levels. The affected children appeared to "catch up" to controls at 12 years of age. Children born seven to twelve years after maternal exposure experienced mildly delayed development, but no differences in behaviour. Effects observed in these children is likely a result of the persistence of PCBs in the human body, resulting in prenatal exposure long after the exposure took place.

These effects are consistent with the observations of poorer short term memory functioning in early childhood, in children exposed prenatally by mothers who had high consumption of Lake Michigan sports fish containing PCBs, amongst other POPs.

People exposed in the Yucheng incident had low resistance, and suffered from a variety of infections. Examination during the first year revealed decreased concentrations of IgM and IgA, decreased percentages of total T-cells, active T-cells and helper T-cells, but normal percentages of B-cells and suppressor T-cells; suppression of delayed type response to recalling antigens; enhancement of lymphocyte spontaneous proliferation and an enhancement in lymphoproliferation to certain mitogens. After three years, some, although not all, of the effects had disappeared. Cancer deaths in both male and female workers involved in the manufacture of electrical capacitors were significantly increased. A significant increase in haematological neoplasms and gastrointestinal cancers was observed in male workers. A non-significant increase in lung cancer was observed. The study was, however, limited by the small numbers of deaths.

IARC has concluded that there is limited evidence for the carcinogenicity of PCBs in humans, and there is sufficient evidence in experimental animals. PCBs are therefore classified as probable humans carcinogens (Group 2A).

Animal Exposure

PCBs have a low acute toxicity to laboratory animals, with acute oral LD50 values in rats in the range of 2 to 10 g/kg body weight. Effects are manifested primarily through chronic exposure. Effects on the liver, skin, immune system, reproductive system, gastrointestinal tract and thyroid gland have been observed and associated with exposure to PCB mixtures or individual congeners.

Adverse reproductive effects observed in several studies on monkeys exposed to PCBs include low birth weights, skin hyperpigmentation, behavioural disturbances, atrophy of the thymus and lymph nodes, bone marrow hypoplasia, and hyperplasia of the gastric mucosa. Female Rhesus Monkeys fed diets containing Aroclor 1016 in the diet were bred after 7 months of dietary exposure. Neonatal weights in the 1.0 ppm group were significantly decreased.

PCBs have not been observed to be teratogenic in studies involving rats and non-human primates when tested orally, during critical periods of organogenesis. A moderate but statistically significant inhibitory effect on the immune system of rhesus monkeys has been observed, resulting from chronic, low level exposure to Aroclor 1254 and that these effects may be due to altered T-cell and/or macrophage function. PCBs have a low acute toxicity to birds, with 5-day dietary LC50 values in the range of 747 mg/kg diet in quail to >5,000 mg/kg in several species. Broiler breeder and leghorn hens who were fed diets Aroclor 1242 for one week experienced reduced hatchability and the effects continued after exposure was terminated.

There is growing evidence linking persistent halogenated aromatic hydrocarbons such as PCBs to reproductive and immunotoxic effects in wildlife. Two groups of 12 female seals (Phoca vitulina) were fed diets of fish from the western part of the Wadden Sea, or from the north-east Atlantic. Residue analysis showed statistically significant differences between the two diets for PCBs and DDE. The average daily intake for group 1 was 1.5 mg PCBs and 0.4 mg DDE, and 0.22 mg and 0.13 mg for group 2. Females were mated with undosed males and reproductive success was significantly lower in group 1.

Mink fed Lake Michigan Coho salmon containing between 10 and 15 ppm PCBs as 30% of their diet for five months failed to whelp as did those fed a diet containing 5 ppm Aroclor 1254. The clinical signs and lesions observed in mink fed a diet containing Lake Michigan coho salmon included anorexia, bloody stools, fatty liver, kidney degeneration and gastric ulcers, and were similar to those fed a diet supplemented with PCBs.

Aquatic Organism Exposure

PCBs are toxic to aquatic organisms, with 96-hour LC50 values in the range of 0.015 mg/L in fathead minnows to 2.74 mg/L in bluegills. Fathead minnows were exposed to Aroclor 1242, 1248 or 1254 in a continuous flow bioassay for 9 months. Reproduction occurred at and below 5.4 μg Aroclor 1242/L, however, results were highly variable.

A significant reduction in spawning was observed in fish exposed to 1.8 μg Aroclor 1254/L. Early life stages of fish are more sensitive to the effects of dioxins, furans, and PCBs. Parts per trillion concentrations of these structurally related chemicals in Lake Trout and Rainbow Trout eggs produce toxicity through sac fry mortality associated with yolk sac edema and haemorrhages.

Bioaccumulation

The degradation of PCBs in the environment depends largely on the degree of chlorination of the biphenyl, with persistence increasing as the degree of chlorination increases. Half-lives for PCBs undergoing photodegradation range from approximately 10 days for a monochlorobiphenyl to 1.5 years for a heptachlorobiphenyl. The persistence of PCBs, combined with the high partition coefficients of various isomers (log K_{ow} ranging from 4.3 to 8.26) provide the necessary conditions for PCBs to bioaccumulate in organisms. Bioconcentration factors of 120,000 and 270,000 have been reported in fathead minnows. Concentration factors in fish exposed to PCBs in their diet were lower than those for fish exposed to PCBs in water, suggesting that PCBs are bioconcentrated (taken up directly from the water) as opposed to being bioaccumulated (taken up by water and in food).

The chemical properties of PCBs (low water solubility, high stability, and semi-volatility) favour their long range transport, and PCBs have been detected in Arctic air, water and organisms.

The main source of PCB exposure to the general population is through food, especially fish. PCB residues were detected in 8.5% of samples, with a maximum of 0.30 mg/kg fat, taken during a survey of the fat of domestic farm animals in Ontario, Canada between 1986 and 1988.

In a survey of foods in Vietnam, the highest levels of PCBs were detected in fish and shellfish, with levels of 760 and 1,400 μg/g fat. The main sources of PCBs in the

Vietnamese diets is cereals (including rice) and vegetables, and the daily intake of 3.7 µg/person/day is comparable to those of some industrialized countries.

DICHLORODIPHENYLTRICHLOROETHANE

DDT was widely used during the Second World War to protect the troops and civilians from the spread of malaria, typhus and other vector borne diseases. After the war, DDT was widely used on a variety of agricultural crops and for the control of disease vectors as well. It is still being produced and used for vector control. Growing concern about adverse environmental effects, especially on wild birds, led to severe restrictions and bans in many developed countries in the early 1970s. The largest agricultural use of DDT has been on cotton which accounted for more than 80% of USA use before its ban there in 1972. DDT is still used to control mosquito vectors of malaria in numerous countries.

50 year old DDT found in a Hospital.

DDT is highly insoluble in water and is soluble in most organic solvents. It is semi-volatile and can be expected to partition into the atmosphere as a result. Its presence is ubiquitous in the environment and residues have even been detected in the Arctic. It is lipophilic and partitions readily into the fat of all living organisms and has been demonstrated to bioconcentrate and biomagnify. The breakdown products of DDT, 1,1-dichloro-2,2-bis (4-chlorophenyl)ethane (DDD or TDE) and 1,1-dichloro-2,2bis (4-chlorophenyl) ethylene) (DDE), are also present virtually everywhere in the environment and are more persistent than the parent compound.

The use of DDT has been banned in 34 countries and severely restricted in 34 other countries. The countries that have banned DDT include Argentina, Australia, Bulgaria, Canada, Colombia, Cyprus, Ethiopia, Finland, Hong Kong, Japan, Lebanon, Mozambique, Norway, Switzerland, and the USA. Countries that have severely restricted its use include Belize, Ecuador, the EU, India, Israel, Kenya, Mexico, Panama, and Thailand.

Potential Effects on Humans

DDT has been widely used in large numbers of people who were sprayed directly in programs to combat typhus and in tropical countries to combat malaria. Dermal exposure to DDT has not been associated with illness or irritation in a number of studies. Studies involving human volunteers who ingested DDT for up to 21 months did not result in any observed adverse effects. A non-significant increase in mortality from liver and biliary cancer and a significant increase in mortality from cerebrovascular disease has been observed in workers involved in the production of DDT. There is some evidence to suggest that DDT may be suppressive to the immune system, possibly by depressing humoral immune responses. Perinatal administration of weakly estrogenic pesticides such as DDT produces estrogen-like alterations of reproductive development, and there is also limited data that suggest a possible association between organochlorines, such as DDT and its metabolite DDE, and the risks of breast cancer.

Potential Effects on Plants and Animals

DDT is not acutely toxic to laboratory animals, with acute oral LD50 values in the range of 100 mg/kg body weight for rats to 1,770 mg/kg for rabbits. In a six generation reproduction study in mice, no effect on fertility, gestation, viability, lactation or survival were observed at a dietary level of 25 ppm. A level of 100 ppm produced a slight reduction in lactation and survival in some generations, but not all, and the effect was not progressive. A level of 250 ppm produced clear adverse reproductive effects. In both these and other studies, no evidence of teratogenicity has been observed.

IARC has concluded that while there is inadequate evidence for the carcinogenicity of DDT in humans, there is sufficient evidence in experimental animals. IARC has classified DDT as a possible human carcinogen (Group 2B).

DDT is highly toxic to fish, with 96-hour LC50 values in the range of 0.4 µg/L in shrimp to 42 µg/L in rainbow trout. It also affects fish behaviour. Atlantic salmon exposed to DDT as eggs experienced impaired balance and delayed appearance of normal behaviour patterns. DDT also affects temperature selection in fish. DDT is acutely toxic to birds with acute oral LD50 values in the range of 595 mg/kg body weight in quail to 1,334 mg/kg in pheasant, however it is best known for its adverse effects on reproduction, especially DDE, which causes egg shell thinning in birds with associated significant adverse impact on reproductive success. There is considerable variation in the sensitivity of bird species to this effect, with birds of prey being the most susceptible and showing extensive egg shell thinning in the wild. American kestrels were fed day old cockerels injected with DDE. Residues of DDE in the eggs correlated closely with the dietary DDE concentration and there was a linear relationship between degree of egg shell thinning and the logarithm of the DDE residue in the egg. Data collected in the field has confirmed this trend. DDT (in conjunction with other halogenated aromatic hydrocarbons) has been linked with feminization and altered sex-ratios of Western

Gull populations off the coast of southern California, and Herring Gull populations in the Great Lakes.

Bioconcentration

DDT and related compounds are very persistent in the environment and as much as 50% can remain in the soil 10-15 years after application. This persistence, combined with a high partition coefficient (log KOW = 4.89-6.91) provides the necessary conditions for DDT to bioconcentrate in organisms. Bioconcentration factors of 154,100 and 51,335 have been recorded for fathead minnows and rainbow trout, respectively. It has been suggested that higher accumulations of DDT at higher trophic levels in aquatic systems results from a tendency for organisms to accumulate more DDT directly from the water, rather than by biomagnification. The chemical properties of DDT (low water solubility, high stability and semi-volatility) favour its long range transport and DDT and its metabolites have been detected in Arctic air, water and organisms. DDT has also been detected in virtually all organochlorine monitoring programs and is generally believed to be ubiquitous throughout the global environment.

Accumulation Studies

DDT and its metabolites have been detected in food from all over the world and this route is likely the greatest source of exposure for the general population. DDE was the second most frequently found residue (21%) in a recent survey of domestic animal fats and eggs in Ontario, Canada, with a maximum residue of 0.410 mg/kg. Residues in domestic animals, however, have declined steadily over the past 20 years. In a survey of Spanish meat and meat products, 83% of lamb samples tested contained at least one of the DDT metabolites investigated, with a mean level of 25 ppb. An average of 76.25 ppb p,p'-DDE was detected in fish samples from Egypt. DDT was the most common organochlorine detected in foodstuffs in Vietnam with mean residue concentrations of 3.2 and 2.0 μg/gm fat in meat and fish, respectively. The estimated daily intake of DDT and its metabolites in Vietnam was 19 μg/person/day. Average residues detected in meat and fish in India were 1.0 and 1.1 μg/g fat respectively, with an estimated daily intake of 48 μg/person/day for DDT and its metabolites.

DDT has also been detected in human breast milk. In a general survey of 16 separate compounds in the breast milk of lactating mothers in four remote villages in Papua, New Guinea, DDT was detected in 100% of samples (41), and was one of only two organochlorines detected. DDT has also been detected in the breast milk of Egyptian women, with an average total DDT detected of 57.59 ppb and an estimated daily intake of total DDT for breast feeding infants of 6.90 μg/kg body weight /day. While lower than the acceptable daily intake of 20.0 μg/kg body weight recommended by the Joing FAO/WHO Meeting on Pesticide Residues (JMPR), its continuing presence raises serious concerns regarding potential effects on developing infants.

DIOXINS

Dioxins are a group of chemically-related compounds that are persistent environmental pollutants (POPs). Dioxins are found throughout the world in the environment and they accumulate in the food chain, mainly in the fatty tissue of animals. More than 90% of human exposure is through food, mainly meat and dairy products, fish and shellfish. Many national authorities have programmes in place to monitor the food supply. Dioxins are highly toxic and can cause reproductive and developmental problems, damage the immune system, interfere with hormones and also cause cancer.

Due to the omnipresence of dioxins, all people have background exposure, which is not expected to affect human health. However, due to the highly toxic potential, efforts need to be undertaken to reduce current background exposure. Prevention or reduction of human exposure is best done via source-directed measures, i.e. strict control of industrial processes to reduce formation of dioxins.

Dioxins are environmental pollutants. They belong to the so-called "dirty dozen" - a group of dangerous chemicals known as persistent organic pollutants (POPs). Dioxins are of concern because of their highly toxic potential. Experiments have shown they affect a number of organs and systems.

Once dioxins enter the body, they last a long time because of their chemical stability and their ability to be absorbed by fat tissue, where they are then stored in the body. Their half-life in the body is estimated to be 7 to 11 years. In the environment, dioxins tend to accumulate in the food chain. The higher an animal is in the food chain, the higher the concentration of dioxins.

The chemical name for dioxin is: 2,3,7,8- tetrachlorodibenzo para dioxin (TCDD). The name "dioxins" is often used for the family of structurally and chemically related polychlorinated dibenzo para dioxins (PCDDs) and polychlorinated dibenzofurans (PCDFs). Certain dioxin-like polychlorinated biphenyls (PCBs) with similar toxic properties are also included under the term "dioxins". Some 419 types of dioxin-related compounds have been identified but only about 30 of these are considered to have significant toxicity, with TCDD being the most toxic.

Sources of Dioxin Contamination

Dioxins are mainly by-products of industrial processes but can also result from natural processes, such as volcanic eruptions and forest fires. Dioxins are unwanted by-products of a wide range of manufacturing processes including smelting, chlorine bleaching of paper pulp and the manufacturing of some herbicides and pesticides. In terms of dioxin release into the environment, uncontrolled waste incinerators (solid waste and hospital waste) are often the worst culprits, due to incomplete burning. Technology is available that allows for controlled waste incineration with low dioxin emissions.

Although formation of dioxins is local, environmental distribution is global. Dioxins are found throughout the world in the environment. The highest levels of these compounds are found in some soils, sediments and food, especially dairy products, meat, fish and shellfish. Very low levels are found in plants, water and air.

Extensive stores of PCB-based waste industrial oils, many with high levels of PCDFs, exist throughout the world. Long-term storage and improper disposal of this material may result in dioxin release into the environment and the contamination of human and animal food supplies. PCB-based waste is not easily disposed of without contamination of the environment and human populations. Such material needs to be treated as hazardous waste and is best destroyed by high temperature incineration in specialised facilities.

Dioxin Contamination Incidents

Many countries monitor their food supply for dioxins. This has led to early detection of contamination and has often prevented impact on a larger scale. In many instances dioxin contamination is introduced via contaminated animal feed, e.g. incidences of increased dioxin levels in milk or animal feed were traced back to clay, fat or citrus pulp pellets used in the production of the animal feed.

Some dioxin contamination events have been more significant, with broader implications in many countries.

In late 2008, Ireland recalled many tons of pork meat and pork products when up to 200 times the safe limit of dioxins were detected in samples of pork. This led to one of the largest food recalls related to a chemical contamination. Risk assessments performed by Ireland indicated no public health concern. The contamination was traced back to contaminated feed. In 1999, high levels of dioxins were found in poultry and eggs from Belgium. Subsequently, dioxin-contaminated animal-based food (poultry, eggs, pork) were detected in several other countries. The cause was traced to animal feed contaminated with illegally disposed PCB-based waste industrial oil.

Large amounts of dioxins were released in a serious accident at a chemical factory in Seveso, Italy, in 1976. A cloud of toxic chemicals, including TCDD, was released into the air and eventually contaminated an area of 15 square kilometres where 37 000 people lived.

Extensive studies in the affected population are continuing to determine the long-term human health effects from this incident. TCDD has also been extensively studied for health effects linked to its presence as a contaminant in some batches of the herbicide Agent Orange, which was used as a defoliant during the Vietnam War. A link to certain types of cancers and also to diabetes is still being investigated.

Although all countries can be affected, most contamination cases have been reported in industrialized countries where adequate food contamination monitoring, greater awareness of the hazard and better regulatory controls are available for the detection of dioxin problems. A few cases of intentional human poisoning have also been reported. The most notable incident is the 2004 case of Viktor Yushchenko, President of the Ukraine, whose face was disfigured by chloracne.

Effects of Dioxins on Human Health

Short-term exposure of humans to high levels of dioxins may result in skin lesions, such as chloracne and patchy darkening of the skin, and altered liver function. Long-term exposure is linked to impairment of the immune system, the developing nervous system, the endocrine system and reproductive functions.

Chronic exposure of animals to dioxins has resulted in several types of cancer. TCDD was evaluated by the WHO's International Agency for Research on Cancer (IARC) in 1997 and 2012. Based on animal data and on human epidemiology data, TCDD was classified by IARC as a "known human carcinogen". However, TCDD does not affect genetic material and there is a level of exposure below which cancer risk would be negligible.

Due to the omnipresence of dioxins, all people have background exposure and a certain level of dioxins in the body, leading to the so-called body burden. Current normal background exposure is not expected to affect human health on average. However, due to the high toxic potential of this class of compounds, efforts need to be undertaken to reduce current background exposure.

Sensitive Groups

The developing fetus is most sensitive to dioxin exposure. Newborn, with rapidly developing organ systems, may also be more vulnerable to certain effects. Some people or groups of people may be exposed to higher levels of dioxins because of their diet (such as high consumers of fish in certain parts of the world) or their occupation (such as workers in the pulp and paper industry, in incineration plants, and at hazardous waste sites).

Prevention and Control of Dioxin Exposure

Proper incineration of contaminated material is the best available method of preventing and controlling exposure to dioxins. It can also destroy PCB-based waste oils. The incineration process requires high temperatures, over 850 °C. For the destruction of large amounts of contaminated material, even higher temperatures – 1000 °C or more - are required.

Prevention or reduction of human exposure is best done via source-directed measures, i.e. strict control of industrial processes to reduce formation of dioxins as much as possible. This is the responsibility of national governments. The Codex Alimentarius Commission adopted a Code of Practice for Source Directed Measures to Reduce Contamination of Foods with Chemicals. Later in 2006 a Code of Practice for the Prevention and Reduction of Dioxin and Dioxin-like PCB Contamination in Food and Feeds was adopted. More than 90% of human exposure to dioxins is through the food supply, mainly meat and dairy products, fish and shellfish. Therefore, protecting the food supply is critical. In addition to source-directed measures to reduce dioxin emissions, secondary contamination of the food supply needs to be avoided throughout the food chain. Good controls and practices during primary production, processing, distribution and sale are all essential in the production of safe food. As indicated through the examples listed above, contaminated animal feed is often the root-cause of food contamination.

Food and feed contamination monitoring systems must be in place to ensure that tolerance levels are not exceeded. It is the responsibility of feed and food producers to assure safe raw materials and safe processes during production, and it is the role of national governments to monitor the safety of food supply and to take action to protect public health. When contamination is suspected, countries should have contingency plans to identify, detain and dispose of contaminated feed and food. The affected population should be examined in terms of exposure (for example, measuring the contaminants in blood or human milk) and effects (for example, clinical surveillance to detect signs of ill health).

Trimming fat from meat and consuming low fat dairy products may decrease the exposure to dioxin compounds. Also, a balanced diet (including adequate amounts of fruits, vegetables and cereals) will help to avoid excessive exposure from a single source. This is a long-term strategy to reduce body burdens and is probably most relevant for girls and young women to reduce exposure of the developing fetus and when breastfeeding infants later on in life. However, the possibility for consumers to reduce their own exposure is somewhat limited.

Identifying Dioxins

The quantitative chemical analysis of dioxins requires sophisticated methods that are available only in a limited number of laboratories around the world. The analysis costs are very high and vary according to the type of sample, but range from over US$ 1000 for the analysis of a single biological sample to several thousand US dollars for the comprehensive assessment of release from a waste incinerator.

Increasingly, biological (cell or antibody) based screening methods are being developed, and theuse of such methods for food and feed samples is increasingly being validated. Such screening methods allow more analyses at a lower cost, and in case of a positive screening test, confirmation of results must be carried out by more complex chemical analysis.

α-HEXACHLOROCYCLOHEXANE

Alpha-HCH is one of the five stable isomers of technical HCH, a pesticide formerly used in agriculture. In general, HCHs are among the most studied pesticides with respect to their environmental fate and effects.

Production

Alpha-HCH by itself is not intentionally produced or available on the market. Instead, alphaHCH is produced as a part of technical HCH which is used as an insecticide or as an intermediate chemical in the manufacturing of lindane (enriched HCH). Currently no production data on technical HCH have been reported, whereas manufacture of lindane still takes place.

HCH is manufactured by photochemical chlorination of benzene which leads to the formation of mainly five stable HCH isomers. The yields of different isomers vary due to technical differences in the production process. The reported ranges are:

- Alpha-HCH: 55-80%,
- Beta-HCH: 5-14%,
- Gamma-HCH: 8-15%,
- Delta-HCH: 6-10%; and
- Epsilon-HCH: 1-5%.

Human Health Impacts

In humans, breathing toxic amounts of alphaHCH can result in blood disorders, dizziness, headaches, and possible changes in the levels of sex hormones in the blood. These effects have occurred in workers exposed to HCH vapors during pesticide manufacturing. People who have swallowed large amounts have had seizures; some have died. Animals that have been fed alphaHCH have had convulsions. All HCH isomers can produce liver and kidney effects. HCH isomers are changed by the body into other chemical products, some of which may be responsible for the harmful effects. Long-term oral administration of alphaHCH or technical-grade HCH to laboratory rodents has been reported to result in liver cancer.

The US Department of Health and Human Services (DHHS) has determined that HCH (all isomers) may reasonably be anticipated to cause cancer in humans. The International Agency for Research on Cancer (IARC) has classified HCH (all isomers) as possibly carcinogenic to humans. The US-EPA has classified alphaHCH as a probable human carcinogen.

OCTABROMODIPHENYL ETHER

Commercial octabromodiphenyl ether (c-octaBDE) is a mixture of several polybrominated diphenyl ethers and congeners. These synthetic brominated compounds have mainly been used as flame retardants. In addition to octaBDE isomers, c-octaBDE contains significant amounts of other component groups (such as pentabromodiphenyl (pentaBDE) and hexabromodiphenyl ethers) with persistent organic pollutant (POP) characteristics.

Adverse Effects

The available ecotoxicity data for the c-octaBDE product show little or no effect on aquatic organisms (short-term fish study and a longer-term Daphnia magna study), sediment organisms (Lumbriculus variegatus) and soil organisms (three species of plant and earthworms Eisenia fetida). However, the EU Risk Assessment Report identifies a risk of secondary poisoning in other species (via ingestion of earthworms) for the hexabromodiphenyl ether component in the c-octaBDE product (from use in polymer applications). Within the EU, c-octaBDE has been classified as "Toxic", due to its effects on human health, with the risk phrases "may cause harm to unborn child", and "possible risk of impaired fertility".

The presence of lower brominated diphenyl ethers in the c-octaBDE products is of concern also from the human health point of view as they are likely to have a higher potential to cause adverse effects.

Persistence

OctaBDE has been found to photodegrade rapidly in a mixture of organic solvents, with a half life of around 5 hours, but the environmental significance of such a finding is uncertain. Besides, octaBDE is predicted to adsorb strongly onto sediment and soil, which means that only a fraction of this PBDE will be exposed to sunlight, thus having the potential to photodegrade. No information is available on the hydrolysis of octaBDE, but it is not expected to be an important process for octaBDE in the environment.

Regarding biotic degradation, octaBDE is not readily biodegradable in standard tests (no degradation seen over 28 days) and is not expected (by analogy with other brominated diphenyl ethers) to degrade rapidly under anaerobic conditions. Nevertheless, other more highly brominated congeners (deca and nonabromodiphenyl ether) have been found to degrade anaerobically in sewage sludge, although at a very slow rate. The evidence seems to indicate that there is little significant biotic or abiotic degradation of octaBDE.

It's worth noting that degradation of polybrominated diphenyl ethers (PBDEs) can yield byproducts that are lower brominated congeners. For instance, Ahn et al. showed that decaBDE immobilised on specific soil/sediment and mineral aerosols yielded a number of penta to triBDEs, via octaBDE as an intermediate step. This may pose an additional environmental concern, as these lower brominated diphenyl ethers are usually more toxic and much more bioaccumulative.

LINDANE

Lindane is an organochlorine chemical and an isomer of hexachlorocyclohexane that has been used both as an agricultural insecticide and as a pharmaceutical treatment for lice and scabies.

Lindane is a neurotoxin that interferes with GABA neurotransmitter function by interacting with the $GABA_A$ receptor-chloride channel complex at the picrotoxin binding site. In humans, lindane affects the nervous system, liver, and kidneys, and may well be a carcinogen. Whether lindane is an endocrine disruptor is unclear.

The World Health Organization classifies lindane as "moderately hazardous", and its international trade is restricted and regulated under the Rotterdam Convention on Prior Informed Consent. In 2009, the production and agricultural use of lindane was banned under the Stockholm Convention on persistent organic pollutants. A specific exemption to that ban allows it to continue to be used as a second-line pharmaceutical treatment for lice and scabies.

Human Health Effects

The EPA and WHO both classify lindane as "moderately" acutely toxic. It has an oral LD_{50} of 88 mg/kg in rats and a dermal LD_{50} of 1000 mg/kg. Most of the adverse human health effects reported for lindane have been related to agricultural uses and chronic, occupational exposure of seed-treatment workers.

Exposure to large amounts of lindane can harm the nervous system, producing a range of symptoms from headache and dizziness to seizures, convulsions, and more rarely, death. Lindane has not been shown to affect the immune system in humans, and it is not considered to be genotoxic. Prenatal exposure to β-HCH, an isomer of lindane and production byproduct, has been associated with altered thyroid hormone levels and could affect brain development.

The Occupational Safety and Health Administration and National Institute for Occupational Safety and Health have set occupational exposure limits (permissible exposure and recommended exposure, respectively) for lindane at 0.5 mg/m^3 at a time-weighted average of eight hours for skin exposure. People can be exposed to lindane in the workplace by inhaling it, absorbing it through their skin, swallowing it, and eye contact. At levels of 50 mg/m^3, lindane is immediately dangerous to life and health.

It is classified as an extremely hazardous substance in the United States, and is subject to strict reporting requirements by facilities which produce, store, or use it in significant quantities.

Cancer Risk

Based primarily on evidence from animal studies, most evaluations of lindane have concluded that it may possibly cause cancer. In 2015, the International Agency for Research on Cancer classified lindane as a known human carcinogen, and in 2001 the EPA concluded there was "suggestive evidence of carcinogenicity, but not sufficient to assess human carcinogenic potential." The U.S. Department of Health and Human Services determined that all isomers of hexachlorocyclohexane, including lindane, "may reasonably be anticipated to cause cancer in humans," and in 1999, the EPA characterized the evidence carcinogenicity for lindane as "suggestive of carcinogenicity, but not sufficient to assess human carcinogenic potential." Lindane and its isomers have also been on California's Proposition 65 list of known carcinogens since 1989. In contrast, the World Health Organization concluded in 2004 that "lindane is not likely to pose a carcinogenic risk to humans." India's BIS considers Lindane a "confirmed carcinogen".

Adverse Reactions

A variety of adverse reactions to lindane pharmaceuticals have been reported, ranging from skin irritation to seizures, and, in rare instances, death. The most common side effects are burning sensations, itching, dryness, and rash. While serious effects are rare and have most often resulted from misuse, adverse reactions have occurred when used properly. The FDA, therefore, requires a so-called black box warning on lindane products, which explains the risks of lindane products and their proper use.

The black box warning emphasizes that lindane should not be used on premature infants and individuals with known uncontrolled seizure disorders, and should be used with caution in infants, children, the elderly, and individuals with other skin conditions (e.g., dermatitis, psoriasis) and people who weigh less than 110 lb (50 kg), as they may be at risk of serious neurotoxicity.

Environmental Contamination

Lindane is a persistent organic pollutant: it is relatively long-lived in the environment, it is transported long distances by natural processes like global distillation, and it can bioaccumulate in food chains, though it is rapidly eliminated when exposure is discontinued.

The production and agricultural use of lindane are the primary causes of environmental contamination, and levels of lindane in the environment have been decreasing in the U.S., consistent with decreasing agricultural usage patterns. The production of lindane generates large amounts of waste hexachlorocyclohexane isomers, and "every ton of lindane manufactured produces about nine tons of toxic waste." Modern manufacturing standards for lindane involve the treatment and conversion of waste isomers to less toxic molecules, a process known as "cracking".

When lindane is used in agriculture, an estimated 12–30% of it volatilizes into the atmosphere, where it is subject to long-range transport and can be deposited by rainfall. Lindane in soil can leach to surface and even ground water, and can bioaccumulate in the food chain. However, biotransformation and elimination are relatively rapid when exposure is discontinued. Most exposure of the general population to lindane has resulted from agricultural uses and the intake of foods, such as produce, meats, and milk, produced from treated agricultural commodities. Human exposure has decreased significantly since the cancellation of agricultural uses in 2006. Even so, the CDC published in 2005 its Third National Report on Human Exposures to Environmental Chemicals, which found no detectable amounts of lindane in human blood taken from a random sampling of about 5,000 people in the US as part of the NHANES study. The lack of detection of lindane in this large human "biomonitoring" study likely reflects the increasingly limited agricultural uses of lindane over the last two decades. The cancellation of agricultural uses in the United States will further reduce the amount of lindane introduced into the environment by more than 99%.

Over time, lindane is broken down in soil, sediment, and water into less harmful substances by algae, fungi, and bacteria; however, the process is relatively slow and dependent on ambient environmental conditions. Lindane residues in honey and beeswax are reported to be the highest of any historical or current pesticide and to continue to pose a threat to honeybee health. The ecological impact of lindane's environmental persistence continues to be debated.

The US EPA determined in 2002 that the agency does not believe that lindane contaminates drinking water in excess of levels considered safe. U.S. Geological Survey teams concluded the same in 1999 and 2000. With regard to lindane medications, the EPA conducted "down-the-drain" estimates of the amount of lindane reaching public water supplies and concluded that lindane levels from pharmaceutical sources were "extremely low" and not of concern.

The EPA has set the maximum contaminant level or "MCL" for lindane allowed in public water supplies and considered safe for drinking at 200 parts per trillion (ppt). By comparison, the state of California imposes a lower MCL for lindane of 19 ppt. However, the California standard is based on a dated 1988 national water criterion that was subsequently revised by the EPA in 2003 to 980 ppt. The EPA stated that the change resulted from "significant scientific advances made in the last two decades particularly in the areas of cancer and noncancer risk assessments." While the EPA considered raising the MCL standard for lindane to 980 ppt at that time, the change was never implemented because states had little difficulty in maintaining lindane levels below the 200 ppt MCL limit already in place. Today, the legally enforceable MCL standard for lindane is 200 ppt, while the national water criterion for lindane is 980 ppt.

Isomers

Lindane is the gamma isomer of hexachlorocyclohexane ("γ-HCH"). In addition to the issue of lindane pollution, some concerns are related to the other isomers of HCH, namely alpha-HCH and beta-HCH, which are notably more toxic than lindane, lack its insecticidal properties, and are byproducts of lindane production. In the 1940s and 1950s, lindane producers stockpiled these isomers in open heaps, which led to ground and water contamination. The International HCH and Pesticide Forum has since been established to bring together experts to address the clean-up and containment of these sites. Modern manufacturing standards for lindane involve the treatment and conversion of waste isomers to less toxic industrial chemicals, a process known as "cracking". Today, only a few production plants remain active worldwide to accommodate public-health uses of lindane and declining agricultural needs. Lindane has not been manufactured in the U.S. since the mid-1970s, but continues to be imported.

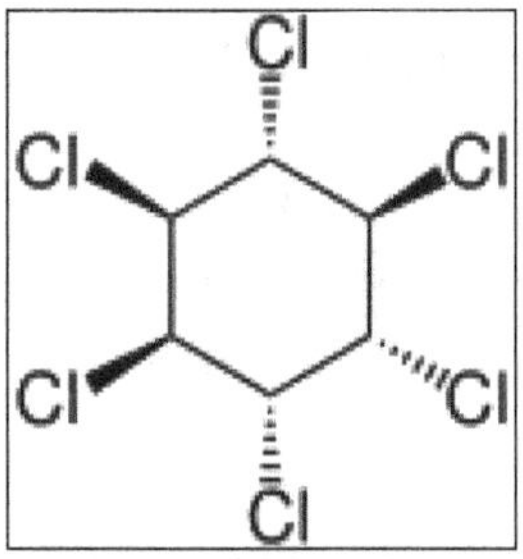

α-hexachlorocyclohexane.

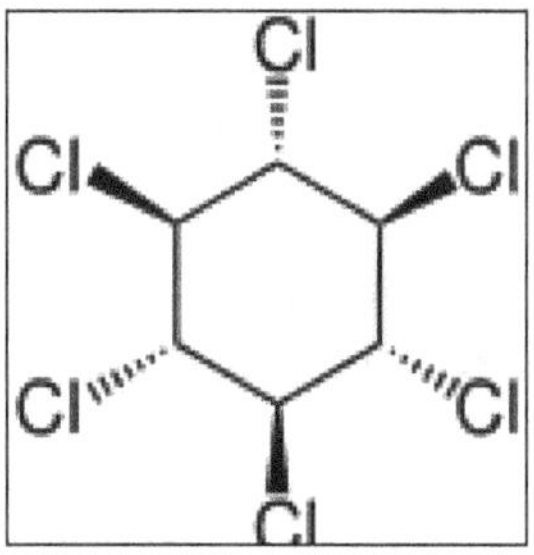

β-hexachlorocyclohexane.

PENTACHLOROBENZENE

Pentachlorobenzene (PeCB) was previously used for various purposes:

- To reduce the viscosity of PCB products employed for heat transfer.
- In dyestuff carriers.
- As a fungicide.
- In a flame retardant.
- As a chemical intermediate such as the production of quintozene (a soil fungicide).

PeCB is also produced unintentionally during combustion in thermal and industrial processes. It appears as an impurity in products such as solvents or pesticides.

PeCB may be emitted to the environment indirectly:

- As a result of waste incineration and barrel burning of household waste.
- In waste streams from pulp and paper mills, iron and steel mills, and petroleum refineries.
- In activated sludge from waste water treatment facilities.

Adverse Effects

PeCB is classified within the EU as "Harmful if swallowed" and "Very toxic to aquatic organisms, may cause long-term adverse effects in the aquatic environment.

Suckling pups from mothers fed a dose of 12.5 mg/kg body weight developed tremors 4 - 14 days after birth. At a maternal dose of 6.3 mg/kg body weight this effect did not occur in pups. In another study, pregnant female rats were administered PeCB at levels of 50, 100 and 200 mg/kg body weight daily on day 6 through 15 of gestation. The number of live foetuses was not affected; however, the mean foetal weight was decreased in the highest dose group. Based on the data available, marine and freshwater organisms do not seem to differ significantly in sensitivity to PeCB. Based upon acute and subchronic animal tests, PeCB is probably moderately toxic to humans. Based on experiments with aquatic species, PeCB is toxic for a range of aquatic organisms.

TETRABROMODIPHENYL ETHER

Polybrominated diphenyl ethers or PBDEs, are organobromine compounds that are used as flame retardant. Like other brominated flame retardants, PBDEs have been used in a wide array of products, including building materials, electronics, furnishings, motor vehicles, airplanes, plastics, polyurethane foams, and textiles. They are structurally akin to the polychlorinated biphenyls (PCBs) and other polyhalogenated compounds, consisting of two halogenated aromatic rings. PBDEs are classified according to the average number of bromine atoms in the molecule. The health hazards of these chemicals have attracted increasing scrutiny, and they have been shown to reduce fertility in humans at levels found in households. Their chlorine analogs are polychlorinated diphenyl ethers (PCDEs). Because of their toxicity and persistence, the industrial production of some PBDEs is restricted under the Stockholm Convention, a treaty to control and phase out major persistent organic pollutants (POPs).

Classes of PBDEs

The family of PBDEs consists of 209 possible substances, which are called congeners (PBDE = $C_{12}H_{(10-x)}Br_xO$ ($x = 1, 2, ..., 10 = m + n$)). The number of isomers for mono-, di-, tri-, tetra-, penta-, hexa-, hepta-, octa-, nona-, and decabromodiphenyl ethers are 3, 12, 24, 42, 46, 42, 24, 12, 3 and 1, respectively. In the United States, PBDEs are marketed with the trade names DE-60F, DE-61, DE-62, and DE-71 applied to pentaBDE mixtures, DE-79 applied to octaBDE mixtures, and DE 83R and Saytex 102E applied to decaBDE mixtures. The available commercial PBDE products are not single compounds or even single congeners but rather mixtures of congeners.

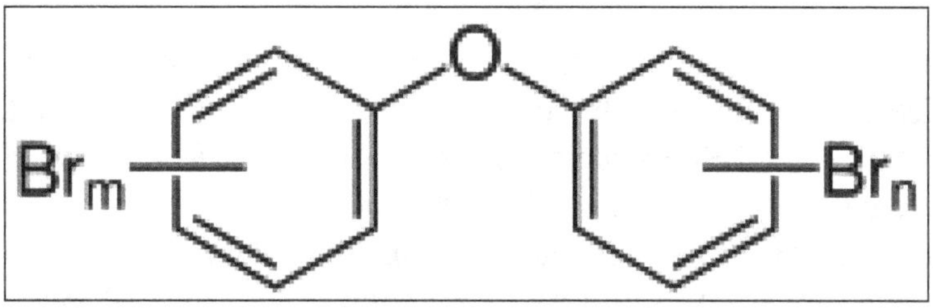

Chemical structure of PBDEs.

Lower Brominated PBDEs

These species average 1–4 bromine atoms per molecule and are regarded as more dangerous because they more efficiently bioaccumulate. Lower-brominated PBDEs have been known to affect hormone levels in the thyroid gland. Studies have linked them to reproductive and neurological risks at certain concentrations or higher.

Higher Brominated PBDEs

These species have average 5 or more bromine atoms per molecule. The commercial mixture, named pentabromodiphenyl ether contains the pentabromo derivative predominantly (50–62%); however, the mixture also contains tetrabromides (24–38%) and hexabromides (4–8%), as well as traces of the tribromides (0–1%). In similar manner, commercial octabromodiphenyl ether is a mixture of homologs: hexa-, hepta-, octa-, nona-, and decabromides.

Health and Environmental Concerns

Since the 1990s, environmental concerns were raised because of the high hydrophobicity of PBDEs and their high resistance to degradation processes. While biodegradation is not considered the main pathway for PBDEs, the photolysis and pyrolysis can be of interest in studies of transformation of PBDEs. People are exposed to low levels of PBDEs through ingestion of food and by inhalation. PBDEs bioaccumulate in blood, breast milk, and fat tissues. Personnel associated with the manufacture of PBDE-containing products are exposed to highest levels of PBDEs. Bioaccumulation is of particular concern in such instances, especially for personnel in recycling and repair plants working on PBDE-containing products. People are also exposed to these chemicals in their domestic environment because of their prevalence in common household items. Studies in Canada have found significant concentrations of PBDEs in common foods such as salmon, ground beef, butter, and cheese. PBDEs have also been found at high levels in indoor dust, sewage sludge, and effluents from wastewater treatment plants. Increasing PBDE levels have been detected in the blood of marine mammals such as harbor seals.

There is also growing concern that PBDEs share the environmental long life and bioaccumulation properties of polychlorinated dibenzodioxins.

It is not known if PBDEs can cause cancer in people, although liver tumors developed in rats and mice that ate extremely large amounts of decaBDE throughout their lifetime. Lower-brominated PBDEs have not yet been tested for cancer in animals. The International Agency for Research on Cancer (IARC) has classified PBDE as a Group 3 carcinogen (not classifiable as to its carcinogenicity to humans) based on inadequate evidence of carcinogenicity in humans and inadequate or limited evidence in experimental animals. The EPA assigns the cancer category Group D (not classifiable as to human carcinogenicity) to mono-, di-, tri-, tetra-, penta-, hexa-, octa-, and nonaBDEs and reports "inadequate information" to classify the specific congeners 2,2',4,4'-tetraBDE, 2,2',4,4',5-pentaBDE, and 2,2',4,4',5,5'-hexaBDE. However, EPA assigns a classification of "suggestive evidence of carcinogenic potential" for decaBDE. The Department of Health and Human Services has not classified PBDEs as carcinogens. The American Conference of Governmental Industrial Hygienists (ACGIH) has no data regarding cancer classifications for PBDEs.

PBDEs have been shown to have hormone-disrupting effects, in particular, on estrogen and thyroid hormones. A 2009 animal study conducted by the US Environmental Protection Agency (EPA) demonstrates that deiodination, active transport, sulfation, and glucuronidation may be involved in disruption of thyroid homeostasis after perinatal exposure to PBDEs during critical developmental time points *in utero* and shortly after birth. The adverse effects on hepatic mechanism of thyroid hormone disruption during development have been shown to persist into adulthood. The EPA noted that PBDEs are particularly toxic to the developing brains of animals. Peer-reviewed studies have shown that even a single dose administered to mice during development of the brain can cause permanent changes in behavior, including hyperactivity.

Swedish scientists first reported substances related to pentaBDE were accumulating in human breast milk. Studies by the Swedish Society for Nature Conservation found for the first time very high levels of more highly brominated PBDEs (BDE-209) in eggs of peregrine falcons. Two forms of PBDEs, penta- and octaBDE, are no longer manufactured in the United States because of health and safety concerns. Based on a comprehensive risk assessment under the Existing Substances Regulation 793/93/EEC, the European Union has completely banned the use of penta- and octaBDE since 2004. However, both chemicals are still found in furniture and foam items made before the phase-out was completed. The most common PBDEs used in electronics are decaBDE.

DecaBDE is banned in Europe for this use and in some U.S. states. For PBDE, EPA has set reference dose of 7 micrograms per kilogram of body weight, which is "believed to be without appreciable effects". However, Linda Birnbaum, PhD, a senior toxicologist formerly with the EPA (now at NIEHS) notes concern: "What I see is another piece of evidence that supports the fact that levels of these chemicals in children appear to be higher than the levels in their parents; I think this study raises a red flag." A previous study by the Environmental Working Group in 2003 published test results showing that the average level of fire-retardants in breast milk from 20 American mothers was 75 times higher than the average levels measured in Europe.

Increasing levels of PBDEs in the environment may be responsible for the increasing incidence of feline hyperthyroidism. A study in 2007 found PBDE levels in cats 20- to 100-fold greater than median levels in U.S. adults, although it was not adequately powered to establish an association between hyperthyroid cats and serum PBDE levels. Subsequent studies have indeed found such an association.

An experiment conducted at Woods Hole Oceanographic Institution in 2005 showed that the isotopic signature of methoxy-PBDEs found in whale blubber contained carbon-14, the naturally occurring radioactive isotope of carbon. Methoxy-PBDEs are produced by some marine species. If the methoxy-PBDEs in the whale had come from artificial (human-made) sources, they would have contained only stable isotopes of carbon because virtually all PBDEs that are produced artificially use petroleum as the source of carbon; all carbon-14 would have long since completely decayed from that source. The isotopic signatures of the PBDEs themselves were not evaluated. The carbon-14 may instead be in methoxy groups enzymatically added to man-made PBDEs.

A 2010 study found that children with higher concentrations of PBDE congeners 47, 99 and 100 in their umbilical cord blood at birth scored lower on tests of mental and physical development between the ages of one and six. Developmental effects were particularly evident at four years of age, when verbal and full IQ scores were reduced 5.5 to 8.0 points for those with the highest prenatal exposures after correcting for sex, ethnicity, tobacco smoke exposure, and mother's IQ.

Sediment Contamination

Increasing environmental concentrations and changing distributions of PBDEs in sediments of the Clyde River Estuary in Scotland, UK have been assessed. Analysis of six sediment cores each of 1 m depth from Glasgow city to Greenock revealed that total concentrations increased toward the river bed surface (0–10 cm). Amounts of PBDE ranged from 1 to 2,645 μg/kg (dry wt. sediment) with a mean of 287 μg/kg (dry wt. sediment). Down-core PBDE congener profiles showed that higher concentrations were due to elevated levels of BDE-209. The majority of the sediment records clearly showed a change from mainly lower molecular weight BDEs 47,99, 183, 153 at lower depths to

BDE-209 near the surface, a change in congener and homologue group patterns that corresponds to the restrictions of penta- and octaBDE commercial mixtures under EU law in 2004–2006.

PERFLUOROOCTANESULFONIC ACID

PFOS is both intentionally produced and is created through unintended degradation of related anthropogenic chemicals. The current intentional use of PFOS is widespread and found in products such as in electric and electronic parts, firefighting foam, photo imaging, hydraulic fluids and textiles. PFOS are still produced in several countries today.

The historical use of PFOS-related substances in the following applications has been confirmed in the US and the EU:

- Firefighting foams,
- Carpets,
- Leather/apparel,
- Textiles/upholstery,
- Paper and packaging,
- Coatings and coating additives,
- Industrial and household cleaning products,
- Pesticides and insecticides.

In the UK study, detailed information has been received from the following sectors that currently use PFOS-related substances:

- Use of existing firefighting foam stock,
- Photographic industry,
- Photolithography and semiconductor,
- Hydraulic fluids,
- Metal plating.

Persistence

PFOS is extremely persistent. It has not showed any degradation in tests of hydrolysis, photolysis or biodegradation at any environmental condition tested. The only known condition whereby PFOS is degraded is through high temperature incineration.

Bioaccumulation

PFOS fulfils the criteria for bioaccumulation based on the much higher concentrations of PFOS in top predators, such as the polar bear, seal, bald eagle and mink (in the Arctic, the US and Sweden), than at lower trophic levels.

Toxicity

The UK and Sweden have proposed the following classification for PFOS in EU:

- Toxic,
- Limited evidence of carcinogenic effect,
- Danger of serious damage to health by prolonged exposure if swallowed,
- May cause harm to the unborn child,
- Toxic to aquatic organisms,
- May cause long-term adverse effects in the aquatic environment.

PFOS has demonstrated toxicity towards mammals in subchronic repeated dose studies at low concentrations, as well as rat reproductive toxicity with mortality of pups occurring shortly after birth. PFOS is toxic to aquatic organisms with mysid shrimp and Chironomus tentans being the most sensitive organisms.

HEXABROMOCYCLODODECANE

Hexabromocyclododecane (HBCD or HBCDD) is a brominated flame retardant. It consists of twelve carbon, eighteen hydrogen, and six bromine atoms tied to the ring. Its primary application is in extruded (XPS) and expanded (EPS) polystyrene foam that is used as thermal insulation in the building industry. Other uses are upholstered furniture, automobile interior textiles, car cushions and insulation blocks in trucks, packaging material, video cassette recorder housing and electric and electronic equipment. According to UNEP, "HBCD is produced in China, Europe, Japan, and the USA. The known current annual production is approximately 28,000 tonnes per year. The main share of the market volume is used in Europe and China". Due to its persistence, toxicity, and ecotoxicity, the Stockholm Convention on Persistent Organic Pollutants decided in May 2013 to list hexabromocyclododecane in Annex A to the Convention with specific exemptions for production and use in expanded polystyrene and extruded polystyrene in buildings. Because HBCD has 16 possible stereo-isomers with different biological activities, the substance poses a difficult problem for manufacture and regulation.

Toxicity

HBCD's toxicity and its harm to the environment are current sources of concern. HBCD can be found in environmental samples such as birds, mammals, fish and other aquatic organisms as well as soil and sediment. On this basis, on 28 October 2008 the European Chemicals Agency decided to include HBCD in the SVHC list, Substances of Very High Concern, within the Registration, Evaluation, Authorisation and Restriction of Chemicals framework. On 18 February 2011, HBCD was listed in the Annex XIV of REACH and hence is subject to Authorisation. HBCD can be used until the so-called "sunset date". After that date, only authorised applications will be allowed in the EU.

HBCD has been found widely present in biological samples from remote areas and supporting evidences for its classification as Persistent, Bioaccumulative and Toxic (PBT) and undergoes long-range environmental transportation. In July 2012, an EU harmonised classification and labelling for HBCD entered into force. HBCD has been classified as a category 2 for reproductive toxicity. Since August 2010 hexabromocyclododecanes are included in the EPA's List of Chemicals of Concern. On May 2013 the Stockholm Convention on Persistent Organic Pollutants (POPs) decided to include HBCD in the Convention's Annex A for elimination, with specific exemptions for expanded and extruded polystyrene in buildings needed to give countries time to phase-in safer substitutes. HBCD is listed for elimination, but with a specific exemption for expanded polystyrene (EPS) and extruded polystyrene (XPS) in buildings. Countries may choose to use this exemption for up to five years after the request for exemption is submitted. Japan was the first country to implement a ban on the import and production of HBCD effective in May 2014.

Because HBCD has 16 possible stereo-isomers with different biological activities, the substance poses a difficult problem for manufacture and regulation. The HBCD commercial mixture is composed of three main diastereomers denoted as alpha (α-HBCD), beta (β-HBCD) and gamma (γ-HBCD) with traces of others. A series of four published in vivo mice studies were conducted between several federal and academic institutions to characterize the toxicokinetic profiles of individual HBCD stereoisomers. The predominant diastereomer in the HBCD mixture, γ-HBCD, undergoes rapid hepatic metabolism, fecal and urinary elimination, and biological conversion to other diastereomers with a short biological half-life of 1–4 days. After oral exposure to the γ-HBCD diastereomer, β-HBCD was detected in the liver and brain, and α-HBCD and β-HBCD was detected in the fat and feces with multiple novel metabolites identified - monohydroxy-pentabromocyclododecane, monohydroxy-pentabromocyclododecene, dihydroxy-pentabromocyclododecene, and dihydroxy-pentabromocyclododecadiene. In contrast, α-HBCD is more biologically persistent, resistant to metabolism, bioaccumulates in lipid-rich tissues after a 10-day repeated exposure study, and has a longer biological half-life of up to 21 days; only α-HBCD was detected in the liver, brain, fat and feces with no stereoisomerization to γ-HBCD or β-HBCD and low trace levels of four different hydroxylated metabolites were identified. Developing mice had higher HBCD tissue levels than adult mice after exposure to either α-HBCD or γ-HBCD indicating the potential for increased susceptibility of the developing young to HBCD effects. The reported toxicokinetic differences of individual HBCD diastereoisomers have important implications for the extrapolation of toxicological studies of the commercial HBCD mixture to the assessment of human risk.

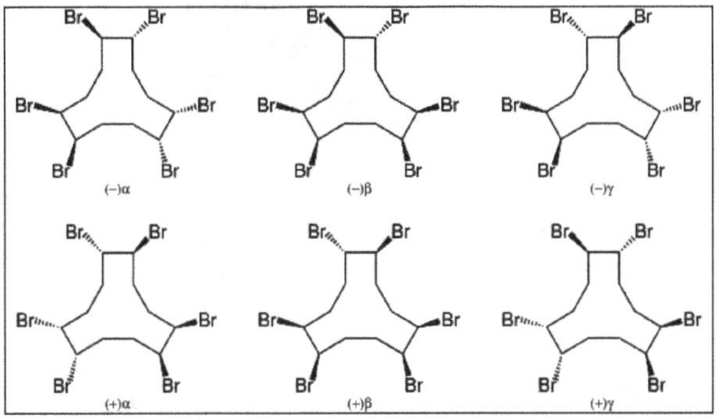

Structures of the six (out of 16 possible) hexabromocyclododecane isomers that are present in the technical product at > 1 %.

Environmental Concerns

Due to its persistence, toxicity, and ecotoxicity, the Stockholm Convention on Persistent Organic Pollutants decided in May 2013 to list hexabromocyclododecane in Annex A to the Convention with specific exemptions for production and use in expanded polystyrene and extruded polystyrene in buildings. Countries may choose to use this exemption for up to five years after the request for exemption is submitted.

There is a large and still increasing stock of HBCD in the anthroposphere, mainly in EPS and XPS insulation boards. A long term environmental monitoring programme run by the Fraunhofer Institute for Molecular Biology and Applied Ecology demonstrates a general trend that HBCD concentrations are decreasing over time. HBCD emissions into the environment are controlled under the voluntary industry emission management programme: the Voluntary Emissions Control Action Programme (VECAP). The VECAP annual report demonstrates continuous decrease of potential emissions of HBCD to the environment.

ENDOSULFAN

Endosulfan is a broad spectrum, non-systemic organochlorine insecticide. It is used to control a number of insects on food crops such as grains, tea, fruits, and vegetables and on non-food crops such as tobacco and cotton. It is also used as a wood preservative.

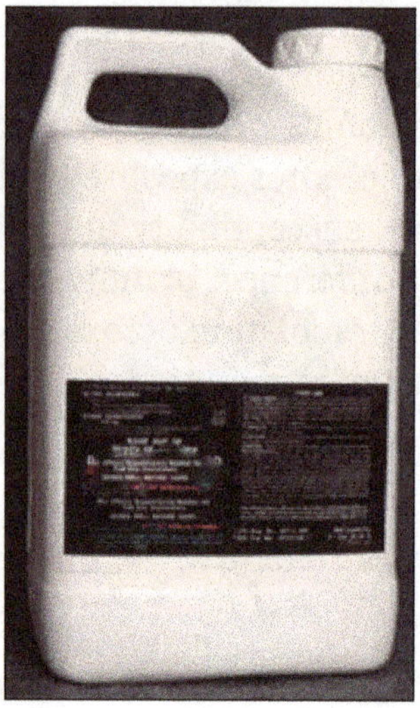

Endosulphan insecticide container.

Endosulfan applied to crops usually breaks down in a few weeks, but Endosulfan preferentially binds to soil particles and may take years to completely break down. Like other POPs chemicals, Endosulfan does not dissolve easily in water and will preferentially bind to aquatic sediments.

Similar to other POPs chemicals, Endosulfan will bioaccumulate into the tissues of animals living in endosulfan-contaminated water.

Potential Effects on Humans

The most likely way for people to be exposed is by eating food contaminated with Endosulfan. Endosulfan has been found in some food products such as oils, fats, and fruit and vegetable products. People have also been exposed to low levels of Endosulfan by skin contact with contaminated soil or by smoking cigarettes made from tobacco that has Endosulfan residues on it. Well water and public water supplies are not likely sources of exposure to Endosulfan. Workers can breathe in the chemical when spraying the pesticide on crops. Exposure can occur by breathing the dust or via contact with skin. Accidental spills and releases to the environment at hazardous waste disposal sites are also possible sources of exposure to endosulfan. The most likely exposure to Endosulfan for people living near hazardous waste sites is through contact with soils containing it.

Endosulfan affects the central nervous system, preventing normal function. Hyperactivity, nausea, dizziness, headache, or convulsions have been observed in adults exposed to high doses. Severe poisoning may result in death.

Studies of the effects of Endosulfan on animals suggest that long-term exposure to Endosulfan can also damage the kidneys, testes, and liver and may possibly affect the body's ability to fight infection. However, it is not known if these effects also occur in humans. It is unknown whether Endosulfan can cause cancer in humans. Studies in animals have provided inconclusive results.

FLAME RETARDANT

The term flame retardants subsumes a diverse group of chemicals which are added to manufactured materials, such as plastics and textiles, and surface finishes and coatings. Flame retardants are activated by the presence of an ignition source and are intended to prevent or slow the further development of ignition by a variety of different physical and chemical methods. They may be added as a copolymer during the polymerisation process, or later added to the polymer at a moulding or extrusion process or (particularly for textiles) applied as a topical finish. Mineral flame retardants are typically additive while organohalogen and organophosphorus compounds can be either reactive or additive.

Classes

Both Reactive and Additive Flame retardants types, can be further separated into several different classes:

- Minerals such as aluminium hydroxide (ATH), magnesium hydroxide (MDH), huntite and hydromagnesite, various hydrates, red phosphorus, and boron compounds, mostly borates.
- Organohalogen compounds: This class includes organochlorines such as chlorendic acid derivatives and chlorinated paraffins; organobromines such as decabromodiphenyl ether (decaBDE), decabromodiphenyl ethane (a replacement for decaBDE), polymeric brominated compounds such as brominated polystyrenes, brominated carbonate oligomers (BCOs), brominated epoxy oligomers (BEOs), tetrabromophthalic anyhydride, tetrabromobisphenol A (TBBPA) and hexabromocyclododecane (HBCD). Most but not all halogenated flame retardants are used in conjunction with a synergist to enhance their efficiency. Antimony trioxide is widely used but other forms of antimony such as the pentoxide and sodium antimonate are also used.
- Organophosphorus compounds: This class includes organophosphates such as triphenyl phosphate (TPP), resorcinol bis(diphenylphosphate) (RDP), bisphenol A diphenyl phosphate (BADP), and tricresyl phosphate (TCP); phosphonates such as dimethyl methylphosphonate (DMMP); and phosphinates such as aluminium diethyl phosphinate. In one important class of flame retardants, compounds contain both phosphorus and a halogen. Such compounds include tris(2,3-dibromopropyl) phosphate (brominated tris) and chlorinated

organophosphates such as tris(1,3-dichloro-2-propyl)phosphate (chlorinated tris or TDCPP) and tetrakis(2-chlorethyl)dichloroisopentyldiphosphate (V6).

- Organic compounds such as carboxylic acid and dicarboxylic acid.

The mineral flame retardants mainly act as additive flame retardants and do not become chemically attached to the surrounding system. Most of the organohalogen and organophosphate compounds also do not react permanently to attach themselves into their surroundings but further work is now underway to graft further chemical groups onto these materials to enable them to become integrated without losing their retardant efficiency. This also will make these materials non emissive into the environment. Certain new non halogenated products, with these reactive and non emissive characteristics have been coming onto the market since 2010, because of the public debate about flame retardant emissions. Some of these new Reactive materials have even received US-EPA approval for their low environmental impacts.

Retardation Mechanisms

The basic mechanisms of flame retardancy vary depending on the specific flame retardant and the substrate. Additive and reactive flame-retardant chemicals can both function in the vapor (gaseous) or condensed (solid) phase.

- Endothermic degradation: Some compounds break down endothermically when subjected to high temperatures. Magnesium and aluminium hydroxides are an example, together with various carbonates and hydrates such as mixtures of huntite and hydromagnesite. The reaction removes heat from the substrate, thereby cooling the material. The use of hydroxides and hydrates is limited by their relatively low decomposition temperature, which limits the maximum processing temperature of the polymers (typically used in polyolefins for wire and cable applications).

- Thermal shielding (solid phase): A way to stop spreading of the flame over the material is to create a thermal insulation barrier between the burning and unburned parts. Intumescent additives are often employed; their role is to turn the polymer surface into a char, which separates the flame from the material and slows the heat transfer to the unburned fuel. Non-halogenated inorganic and organic phosphate flame retardants typically act through this mechanism by generating a polymeric layer of charred phosphoric acid.

- Dilution of gas phase: Inert gases (most often carbon dioxide and water) produced by thermal degradation of some materials act as diluents of the combustible gases, lowering their partial pressures and the partial pressure of oxygen, and slowing the reaction rate.

- Gas phase radical quenching: Chlorinated and brominated materials undergo thermal degradation and release hydrogen chloride and hydrogen bromide or, if used in the presence of a synergist like antimony trioxide, antimony halides. These react with the highly reactive H• and OH• radicals in the flame, resulting in an inactive molecule and a Cl• or Br• radical. The halogen radical is much less reactive compared to H• or OH•, and therefore has much lower potential to propagate the radical oxidation reactions of combustion.

Use and Effectiveness

Fire Safety Standards

Flame retardants are typically added to industrial and consumer products to meet flammability standards for furniture, textiles, electronics, and building products like insulation.

In 1975, California began implementing Technical Bulletin 117 (TB 117), which requires that materials such as polyurethane foam used to fill furniture be able to withstand a small open flame, equivalent to a candle, for at least 12 seconds. In polyurethane foam, furniture manufacturers typically meet TB 117 with additive halogenated organic flame retardants. Although no other U.S. states have a similar standard, because California has such a large market many manufacturers meet TB 117 in products that they distribute across the United States. The proliferation of flame retardants, and especially halogenated organic flame retardants, in furniture across the United States is strongly linked to TB 117.

In response to concerns about the health impacts of flame retardants in upholstered furniture, in February 2013 California proposed modifying TB 117 to require that fabric covering upholstered furniture meet a smolder test and to eliminate the foam flammability standards. Gov. Jerry Brown signed the modified TB117-2013 in November and it became effective in 2014. The modified regulation does not mandate a reduction in flame retardants.

However, these questions of eliminating emissions into the environment from flame retardants can be solved by using a new classification of highly efficient flame retardants, which do not contain halogen compounds, and which can also be keyed permanently into the chemical structure of the foams used in the furniture and bedding industries. The resulting foams have been certified to produce no flame retardant emissions. This new technology is based on entirely newly developed "Green Chemistry" with the final foam containing about one third by weight of natural oils. Use of this technology in the production of California TB 117 foams, would allow continued protection for the consumer against open flame ignition whilst providing the newly recognized and newly needed protection, against chemical emissions into home and office environments. More recent work during 2014 with this "Green Chemistry" has shown that foams containing about fifty percent of natural oils can be made which produce far less smoke when involved in fire situations. The ability of these low emission foams to reduce smoke emissions by up to 80% is an interesting property which will aid escape from fire situations and also lessen the risks for first responders i.e. emergency services in general and fire department personnel in particular.

In Europe, flame retardant standards for furnishings vary, and are their most stringent in the UK and Ireland. Generally the ranking of the various common flame retardant tests worldwide for furniture and soft furnishings would indicate that the California test Cal TB117 - 2013 test is the most straightforward to pass, there is increasing difficulty in passing Cal TB117 -1975 followed by the British test BS 5852 and followed by Cal TB133. One of the most demanding flammability tests worldwide is probably the US Federal Aviation Authority test for aircraft seating which involves the use of a kerosene burner which blasts flame at the test piece. The 2009 Greenstreet Berman study, carried out by the UK government, showed that in the period between 2002 and 2007 the UK Furniture and Furnishings Fire Safety Regulations accounted for 54 fewer deaths per year, 780 fewer non-fatal casualties per year and 1065 fewer fires each year following the introduction of the UK furniture safety regulations in 1988.

Effectiveness

The effectiveness of flame retardant chemicals at reducing the flammability of consumer products in house fires is disputed. Advocates for the flame retardant industry, such as the American Chemistry Council's North American Flame Retardant Alliance, cite a study from the National Bureau of Standards indicating that a room filled with flame-retarded products (a polyurethane foam-padded chair and several other objects, including cabinetry and electronics) offered a 15-fold greater time window for occupants to escape the room than a similar room free of flame retardants. However, critics of this position, including the lead study author, argue that the levels of flame retardant used in the 1988 study, while found commercially, are much higher than the levels required by TB 117 and used broadly in the United States in upholstered furniture.

Another study concluded flame retardants are an effective tool to reduce fire risks without creating toxic emissions.

Several studies in the 1980s tested ignition in whole pieces of furniture with different upholstery and filling types, including different flame retardant formulations. In particular, they looked at maximum heat release and time to maximum heat release, two key indicators of fire danger. These studies found that the type of fabric covering had a large influence on ease of ignition, that cotton fillings were much less flammable than polyurethane foam fillings, and that an interliner material substantially reduced the ease of ignition. They also found that although some flame retardant formulations decreased the ease of ignition, the most basic formulation that met TB 117 had very little effect. In one of the studies, foam fillings that met TB 117 had equivalent ignition times as the same foam fillings without flame retardants. A report from the Proceedings of the Polyurethane Foam Association also showed no benefit in open-flame and cigarette tests with foam cushions treated with flame retardants to meet TB 117. However, other scientists support this open-flame test.

Environmental and Health Issues

The environmental behaviour of flame retardants has been studied since the 1990s. Mainly brominated flame retardants were found in many environmental compartments and organisms including humans, and some individual substances were found to have toxic properties. Therefore, alternatives have been demanded by authorities, NGOs and equipment manufacturers. The EU-funded collaborative research project ENFIRO started out from the assumption that not enough environmental and health data were known of alternatives to the established brominated flame retardants. In order to make the evaluation fully comprehensive, it was decided to compare also material and fire performance as well as attempt a life cycle assessment of a reference product containing halogen free versus brominated flame retardants. About a dozen halogen free flame retardants were studied representing a large variety of applications, from engineering plastics, printed circuit boards, encapsulants to textile and intumescent coatings. A large group of the studied flame retardants were found to have a good environmental and health profile: ammonium polyphosphate (APP), Aluminium diethyl phosphinate (Alpi), aluminium hydroxide (ATH), magnesium hydroxide (MDH), melamine polyphosphate (MPP), dihydrooxaphosphaphenanthrene (DOPO), zinc stannate (ZS) and zinc hydroxstannate (ZHS). Overall, they were found to have a much lower tendency to bioaccumulate in fatty tissue than the studied brominated flame retardants.

The tests on the fire behaviour of materials with different flame retardants revealed that halogen free flame retardants produce less smoke and toxic fire emissions, with the exception of the aryl phosphates RDP and BDP in styrenic polymers. The leaching experiments showed that the nature of the polymer is a dominating factor and that the leaching behaviour of halogen free and brominated flame retardants is comparable. The more porous or "hydrophilic" a polymers is the more flame retardants can be released. However, moulded plates which represent real world plastic products showed much lower leaching levels than extruded polymer granules. The impact assessment studies reconfirmed that the improper waste and recycling treatment of electronic products with brominated flame retardants can produce dioxins which is not the case with halogen free alternatives. Furthermore, the United States Environmental Protection Agency (US-EPA) has been carrying out a series of projects related to the environmental assessment of alternative flame retardants, the "design for environment" projects on flame retardants for printed wiring boards and alternatives to decabromo diphenylethers and hexabromocyclododecane (HBCD).

In 2009, the U.S. National Oceanic and Atmospheric Administration (NOAA) released a report on polybrominated diphenyl ethers (PBDEs) and found that, in contrast to earlier reports, they were found throughout the U.S. coastal zone. This nationwide survey found that New York's Hudson Raritan Estuary had the highest overall concentrations of PBDEs, both in sediments and shellfish. Individual sites with the highest PBDE measurements were found in shellfish taken from Anaheim Bay, California, and four sites in the Hudson Raritan Estuary. Watersheds that include the Southern California Bight, Puget Sound, the central and eastern Gulf of Mexico off the coast of Tampa and St. Petersburg, in Florida, and the waters of Lake Michigan near Chicago and Gary, Indiana, also were found to have high PBDE concentrations.

Health Concerns

The earliest flame retardants, polychlorinated biphenyls (PCBs), were banned in the U.S. in 1977 when it was discovered that they were toxic. Industries used brominated flame retardants instead, but these are now receiving closer scrutiny. In 2004 and 2008 the EU banned several types of polybrominated diphenyl ethers (PBDEs). Negotiations between the EPA and the two U.S. producers of DecaBDE (a flame retardant that has been used in electronics, wire and cable insulation, textiles, automobiles and airplanes, and other applications), Albemarle Corporation and Chemtura Corporation, and the largest U.S. importer, ICL Industrial Products, Inc., resulted in commitments by these companies to phase out decaBDE for most uses in the United States by December 31, 2012, and to end all uses by the end of 2013. The state of California has listed the flame retardant chemical chlorinated Tris (tris(1,3-dichloro-2-propyl) phosphate or TDCPP) as a chemical known to cause cancer. In December 2012, the California nonprofit Center for Environmental Health filed notices of intent to sue several leading retailers and producers of baby products for violating California law for failing to label products containing this cancer-causing flame retardant. While the demand for brominated and chlorinated flame retardants in North America and Western Europe is declining, it is rising in all other regions.

There is a potential association between the exposure to the Phosphorus Flame Retardants (PFR) in residential indoor dust and the development of allergies, asthma and dermatitis. A study was conducted in 2014 by Araki, A. et al. in Japan to assess this relationship. They found a significant association between the Tris (2-chloro-iso-propyl) phosphate (TCIPP) and atopic dermatitis with an odds ratio of 2.43. They also found that the Tributyl phosphate was associated with the development of allergic rhinitis and asthma with an odds ratio of 2.55 & 2.85 respectively.

Nearly all Americans tested have trace levels of flame retardants in their body. Recent research links some of this exposure to dust on television sets, which may have been generated from the heating of the flame retardants in the TV. Careless disposal of TVs and other appliances such as microwaves or old computers may greatly increase the amount of environmental contamination. A recent study conducted by Harley *et al.* 2010 on pregnant women, living in a low-income, predominantly Mexican-immigrant community in California showed a significant decrease in fecundity associated with PBDE exposure in women.

Another study conducted by Chevrier et al. 2010 measured the concentration of 10 PBDE congeners, free thyroxine (T4), total T4, and thyroid-stimulating hormone (TSH) in 270 pregnant women around the 27th week of gestation. Associations between PBDEs and free and total T4 were found to be statistically insignificant. However, authors did find a significant association amongst exposure to PBDEs and lower TSH during pregnancy, which may have implications for maternal health and fetal development.

A prospective, longitudinal cohort study initiated after 11 September 2001, including 329 mothers who delivered in one of three hospitals in lower Manhattan, New York, was conducted by Herbstman *et al.* 2010. Authors of this study analyzed 210 cord blood specimens for selected PBDE congeners and assessed neurodevelopmental effects in the children at 12–48 and 72 months of age. Results showed that children who had higher cord blood concentrations of polybrominated diphenyl ethers (PBDEs) scored lower on tests of mental and motor development at 1–4 and 6 years of age. This was the first study to report any such associations in humans.

A similar study was conducted by Roze et al. 2009 in The Netherlands on 62 mothers and children to estimate associations between 12 Organohalogen compounds (OHCs), including polychlorinated biphenyls (PCBs) and brominated diphenyl ether (PBDE) flame retardants, measured in maternal serum during the 35th week of pregnancy and motor performance (coordination, fine motor skills), cognition (intelligence, visual perception, visuomotor integration, inhibitory control, verbal memory, and attention), and behavior scores at 5–6 years of age. Authors demonstrated for the first time that transplacental transfer of polybrominated flame retardants was associated with the development of children at school age.

Another study was conducted by Rose et al. in 2010 to measure circulating PBDE levels in 100 children between 2 and 5 years of age from California. The PBDE levels according to this study, in 2- to 5-year-old California children was 10 to 1,000 fold higher than European children, 5 times higher than other U.S. children and 2 to 10 times higher than U.S. adults. They also found that diet, indoor environment, and social factors influenced children's body burden levels. Eating poultry and pork contributed to elevated body burdens for nearly all types of flame retardants. Study also found that lower maternal education was independently and significantly associated with higher levels of most flame retardant congeners in the children.

San Antonio Statement on Brominated and Chlorinated Flame Retardants 2010: A group of 145 prominent scientists from 22 countries signed the first-ever consensus statement documenting health hazards from flame retardant chemicals found at high levels in home furniture, electronics, insulation, and other products. This statement documents that, with limited fire safety benefit, these flame retardants can cause serious health issues, and, as types of flame retardants are banned, the alternatives should be proven safe before being used. The group also wants to change widespread policies that require use of flame retardants.

A number of recent studies suggest that dietary intake is one of the main routes to human exposure to PBDEs. In recent years, PBDEs have become widespread environmental pollutants, while body burden in the general population has been increasing. The results do show notable coincidences between the China, Europe, Japan, and United States such as dairy products, fish, and seafood being a cause of human exposure to PBDEs due to the environmental pollutant.

A February 2012 study genetically engineered female mice to have mutations in the x-chromosome MECP2 gene, linked to Rett syndrome, a disorder in humans similar to autism. After exposure to BDE-47 (a PDBE) their offspring, who were also exposed, had lower birth weights and survivability and showed sociability and learning deficits. A January 2013 study of mice showed brain damage from BDP-49, via inhibiting of the mitochondrial ATP production process necessary for brain cells to get energy. Toxicity was at very low levels. The study offers a possible pathway by which PDBEs lead to autism.

Mechanisms of Toxicity

Direct Exposure

Many halogenated flame retardants with aromatic rings, including most brominated flame retardants, are likely thyroid hormone disruptors. The thyroid hormones triiodothyronine (T3) and thyroxine (T4) carry iodine atoms, another halogen, and are structurally similar to many aromatic halogenated flame retardants, including PCBs, TBBPA, and PBDEs. Such flame retardants therefore appear to compete for binding sites in the thyroid system, interfering with normal function of thyroid transport proteins (such as transthyretin) in vitro and thyroid hormone receptors. A 2009 *in vivo* animal study conducted by the US Environmental Protection Agency (EPA) demonstrated that deiodination, active transport, sulfation, and glucuronidation may be involved in disruption of thyroid homeostasis after perinatal exposure to PBDEs during critical developmental time points in utero and shortly after birth. Disruption of deiodinase as reported in the Szabo et al., 2009 *in vivo* study was supported in a follow-up *in vitro* study. The adverse effects on hepatic mechanism of thyroid hormone disruption during development have been shown to persist into adulthood. The EPA noted that PBDEs are particularly toxic to the developing brains of animals. Peer-reviewed studies have shown that even a single dose administered to mice during development of the brain can cause permanent changes in behavior, including hyperactivity.

Based on *in vitro* laboratory studies, several flame retardants, including PBDEs, TBBPA, and BADP, likely also mimic other hormones, including estrogens, progesterone, and androgens. Bisphenol A compounds with lower degrees of bromination seem to exhibit greater estrogenicity. Some halogenated flame retardants, including the less-brominated PBDEs, can be direct neurotoxicants in *in vitro* cell culture studies: By altering calcium homeostasis and signalling in neurons, as well as neurotransmitter release and uptake at synapses, they interfere with normal neurotransmission. Mitochondria may be particularly vulnerable to PBDE toxicity due to their influence on oxidative stress and calcium activity in mitochondria. Exposure to PBDEs can also alter neural cell differentiation and migration during development.

Degradation Products

Many flame retardants degrade into compounds that are also toxic, and in some cases the degradation products may be the primary toxic agent:

- Halogenated compounds with aromatic rings can degrade into dioxins and dioxin-like compounds, particularly when heated, such as during production, a fire, recycling, or exposure to sun. Chlorinated dioxins are among the highly toxic compounds listed by the Stockholm Convention on Persistent Organic Pollutants.
- Polybrominated diphenyl ethers with higher numbers of bromine atoms, such as decaBDE, are less toxic than PBDEs with lower numbers of bromine atoms, such as pentaBDE. However, as the higher-order PBDEs degrade biotically or abiotically, bromine atoms are removed, resulting in more toxic PBDE congeners.
- When some halogenated flame retardants such as PBDEs are metabolized, they form hydroxylated metabolites that can be more toxic than the parent compound. These hydroxylated metabolites, for example, may compete more strongly to bind with transthyretin or other components of the thyroid system, can be more potent estrogen mimics than the parent compound, and can more strongly affect neurotransmitter receptor activity.
- Bisphenol-A diphenyl phosphate (BADP) and tetrabromobisphenol A (TBBPA) likely degrade to bisphenol A (BPA), an endocrine disruptor of concern.

Routes of Exposure

People can be exposed to flame retardants through several routes, including diet; consumer products in the home, vehicle, or workplace; occupation; or environmental contamination near their home or workplace. Residents in North America tend to have substantially higher body levels of flame retardants than people who live in many other developed areas, and around the world human body levels of flame retardants have increased over the last 30 years.

Exposure to PBDEs has been studied the most widely. As PBDEs have been phased out of use due to health concerns, organophosphorus flame retardants, including halogenated organophosphate flame retardants, have frequently been used to replace them. In some studies, indoor air concentrations of phosphorus flame retardants has been found to be greater than indoor air concentrations of PBDEs. The European Food

Safety Authority (EFSA) issued in 2011 scientific opinions on the exposure to HBCD and TBBPA and its derivates in food and concluded that current dietary exposure in the European Union does not raise a health concern

Exposure in the General Population

The body burden of PBDEs in Americans correlates well with the level of PBDEs measured in swabs of their hands, likely picked up from dust. Dust exposure may occur in the home, car, or workplace. Levels of PBDEs can be as much as 20 times higher in vehicle dust as in household dust, and heating of the vehicle interior on hot summer days can break down flame retardants into more toxic degradation products. However, blood serum levels of PBDEs appear to correlate most highly with levels found in dust in the home. 60-80% of exposures are due to dust inhalation or ingestion.. In addition to this, 20% to 40% of adult U.S. exposure to PBDEs is through food intake as PBDEs bioaccumulate in the food chain. High concentration can be found in meat, dairy and fish with the remaining exposure largely due to dust inhalation or ingestion. Individuals can also be exposed through electronic and electrical devices. Young children in the United States tend to carry higher levels of flame retardants per unit body weight than do adults. Infants and toddlers are particularly exposed to halogenated flame retardants found in breast milk and dust. Because many halogenated flame retardants are fat-soluble, they accumulate in fatty areas such as breast tissue and are mobilized into breast milk, delivering high levels of flame retardants to breast-feeding infants. PBDEs also cross the placenta, meaning infants are exposed in utero. Mothers thyroid hormone (T4) level can be disrupted and exposure in utero in rat studies has been demonstrated to alter motor control, delay sensory development and puberty.

Another reason for high levels of exposure in young children are due to aging consumer products age, small particles of material become dust particles in the air and land on surfaces around the home, including the floor. Young children crawling and playing on the floor frequently bring their hands to their mouths, ingesting about twice as much house dust as adults per day in the United States. Children also have a higher food intake per kilogram of bodyweight compared to adults. Young children are also exposed to flame retardants through their clothing, car seats and toys. The introduction of these chemicals came about after the tragic death of children wearing brushed rayon fabric that would ignite easily. The U.S enacted the Flammable Fabrics Act passed in 1953 after which, flame retardants were mandated to be added to many children's items, including pajamas. While flame retardants are shown to decrease the risk of burn injuries in children, the risks of thyroid disruption as well as physical and cognitive developmental delays, are not outweighed.

A study found that gymnasts are exposed to some flame-retardant products such as PentaBDE and TBB more than the general population in the United States. After testing hand-wipe samples before and after the exercise, they found that the BDE-153 concentration was four to over six times greater among gymnasts than the United States population. Also, the PentaBDE concentration was higher up to three times after exercise compared to the level before; indicating a higher level of the flame-retardants on the training equipment. Moreover, they also found several flame-retardant products with different concentrations in the air and dust that were higher in the gym than residencies. However, the study was performed on a small sample size; and further studies are recommended to assess the association.

Occupational Exposure

Some occupations expose workers to higher levels of halogenated flame retardants and their degradation products. A small study of U.S. foam recyclers and carpet installers, who handle padding often made from recycled polyurethane foam, showed elevated levels of flame retardants in their tissues. Workers in electronics recycling plants around the world also have elevated body levels of flame retardants relative to the general population. Environmental controls can substantially reduce this exposure, whereas workers in areas with little oversight can take in very high levels of flame retardants. Electronics recyclers in Guiyu, China, have some of the highest human body levels of PBDEs in the world. A study conducted in Finland determined the occupational exposure of workers to brominated flame retardants and chlorinated flame retardants (TBBPA, PBDEs, DBDPE, HBCD, Hexabromobenzene and Dechlorane plus). In 4 recycling sites of waste electrical and electronic equipment (WEEE), the study concluded that control measures implemented on site significantly reduced the exposure. Workers making products that contain flame retardants (such as vehicles, electronics, and baby products) may be similarly exposed. U.S. firefighters can have elevated levels of PBDEs and high levels of brominated furans, toxic degradation products of brominated flame retardants.

Environmental Exposure

Flame retardants manufactured for use in consumer products have been released into environments around the world. The flame retardant industry has developed a voluntary initiative to reduce emissions to the environment (VECAP) by promoting best practices during the manufacturing process. Communities near electronics factories and disposal facilities, especially areas with little environmental oversight or control, develop high levels of flame retardants in air, soil, water, vegetation, and people.

Organophosphorus flame retardants have been detected in wastewater in Spain and Sweden, and some compounds do not appear to be removed thoroughly during water treatment. Organophosphorus flame-retardants were also found in tap and bottled drinking water in China. Likewise in the Elbe river in Germany.

Disposal

When products with flame retardants reach the end of their usable life, they are typically recycled, incinerated, or landfilled.

Recycling can contaminate workers and communities near recycling plants, as well as new materials, with halogenated flame retardants and their breakdown products. Electronic waste, vehicles, and other products are often melted to recycle their metal components, and such heating can generate toxic dioxins and furans. When wearing Personal Protection Equipment (PPE) and when a ventilation system is installed, exposure of workers to dust can be significantly reduced, as shown in the work conducted by the recycling plant Stena-Technoworld AB in Sweden. Brominated flame retardants may also change the physical properties of plastics, resulting in inferior performance in recycled products and in "downcycling" of the materials. It appears that plastics with brominated flame retardants are mingling with flame-retardant-free plastics in the recycling stream and such downcycling is taking place.

Poor-quality incineration similarly generates and releases high quantities of toxic degradation products. Controlled incineration of materials with halogenated flame retardants, while costly, substantially reduces release of toxic byproducts.

Many products containing halogenated flame retardants are sent to landfills. Additive, as opposed to reactive, flame retardants are not chemically bonded to the base material and leach out more easily. Brominated flame retardants, including PBDEs, have been observed leaching out of landfills in industrial countries, including Canada and South Africa. Some landfill designs allow for leachate capture, which would need to be treated. These designs also degrade with time.

POLYCYCLIC AROMATIC HYDROCARBON

Polycyclic aromatic hydrocarbons are hydrocarbons—organic compounds containing only carbon and hydrogen—that are composed of multiple aromatic rings (organic rings in which the electrons are delocalized). The simplest such chemicals are naphthalene, having two aromatic rings, and the three-ring compounds anthracene and phenanthrene.

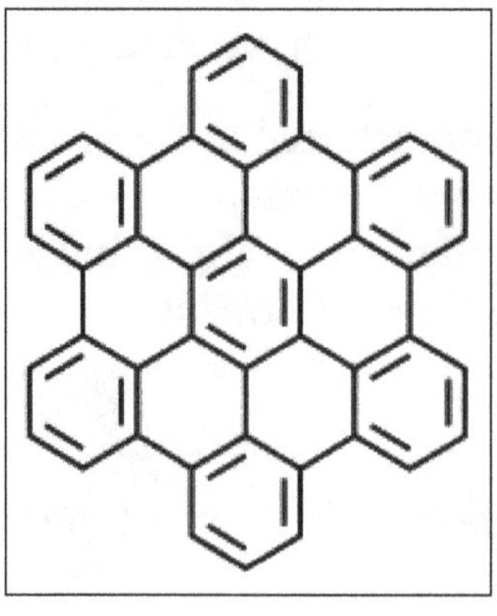

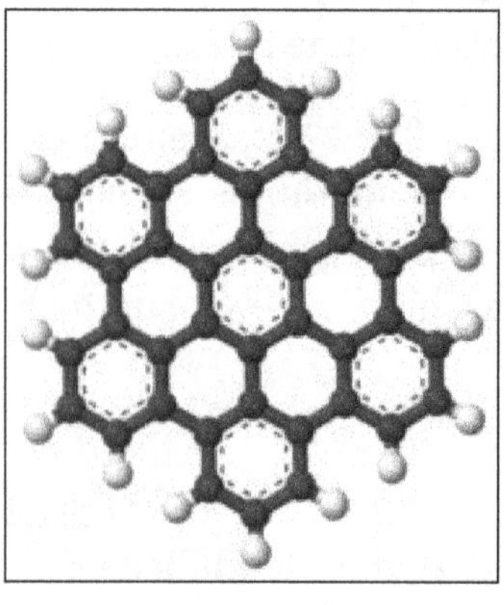

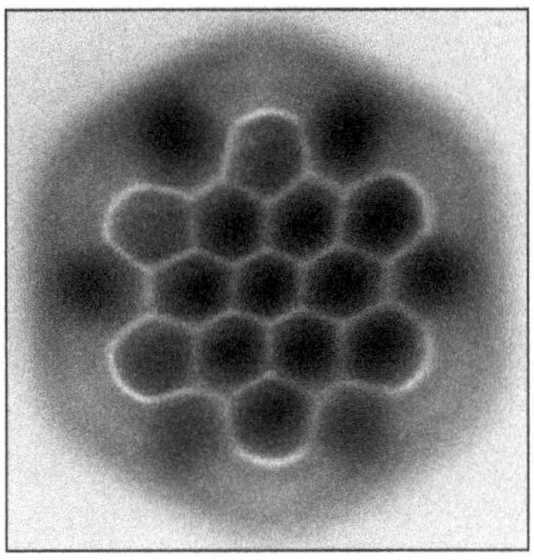

Three representations of hexabenzocoronene, a polycylic aromatic hydrocarbon. Top: standard line-angle schematic, where carbon atoms are represented by the vertices of the hexagons and hydrogen atoms are inferred. Middle: ball-and-stick model showing all carbon and hydrogen atoms.

PAHs are uncharged, non-polar molecules found in coal and in tar deposits. They are also produced by the thermal decomposition of organic matter (for example, in engines and incinerators or when biomass burns in forest fires). PAHs are abundant in the universe, and have recently been found to have formed possibly as early as the first couple of billion years after the Big Bang, in association with formation of new stars and exoplanets. Some studies suggest that PAHs account for a significant percentage of all carbon in the universe.

Polycyclic aromatic hydrocarbons are discussed as possible starting materials for abiotic syntheses of materials required by the earliest forms of life.

Nomenclature, Structure and Properties

Nomenclature and Structure

By definition, polycyclic aromatic hydrocarbons have multiple cycles, precluding benzene from being considered a PAH. Naphthalene is considered the simplest polycyclic aromatic hydrocarbon by the US EPA and US CDC for policy contexts. Other authors consider PAHs to start with the tricyclic species phenanthrene and anthracene.

PAHs are not generally considered to contain heteroatoms or carry substituents.

PAHs with five or six-membered rings are most common. Those composed only of six-membered rings are called *alternant* PAHs, which include *benzenoid* PAHs. The following are examples of PAHs that vary in the number and arrangement of their rings:

- Principal PAH Compounds:
    - Anthracene,
    - Phenanthrene,

- Phenalene,
- Tetracene,
- Chrysene,
- Triphenylene,
- Pyrene,
- Pentacene,
- Benzo[*a*]pyrene,
- Corannulene,
- Benzo[*ghi*]perylene,
- Coronene,
- Ovalene,
- Benzo[*c*]fluorene.

Physicochemical Properties and Bonding

PAHs are nonpolar and lipophilic. Larger PAHs are generally insoluble in water, although some smaller PAHs are soluble and known contaminants in drinking water. The larger members are also poorly soluble in organic solvents and in lipids. They are usually colorless.

The aromaticity varies for PAHs. According to Clar's rule, the resonance structure of a PAH that has the largest number of disjoint aromatic pi sextets–i.e. benzene-like moieties–is the most important for the characterization of the properties of that PAH.

Benzene-substructure resonance analysis for Clar's rule

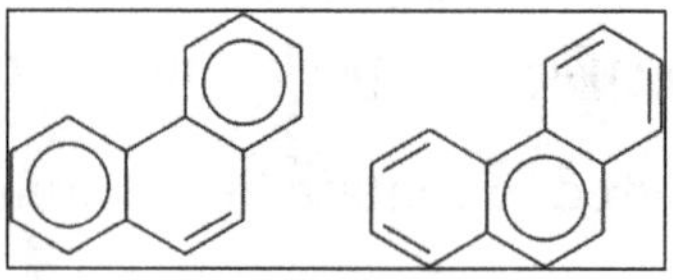

Phenanthrene.

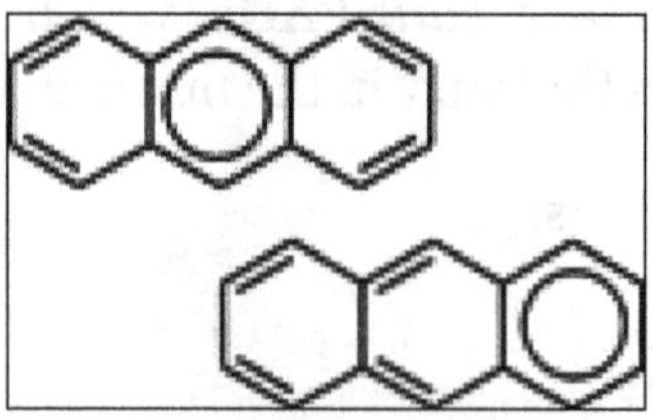

Anthracene.

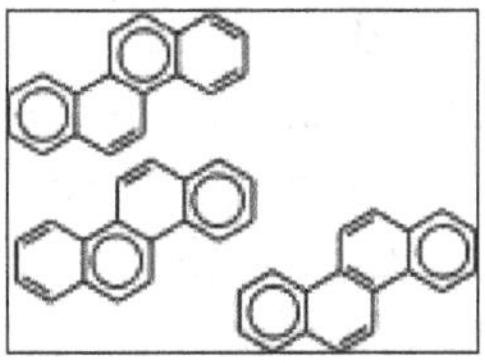

Chrysene

For example, in phenanthrene one Clar structure has two sextets—the first and third rings—while the other resonance structure has just one central sextet; therefore in this molecule the outer rings have greater aromatic character whereas the central ring is less aromatic and therefore more reactive. In contrast, in anthracene the resonance structures have one sextet each, which can be at any of the three rings, and the aromaticity spreads out more evenly across the whole molecule. This difference in number of sextets is reflected in the differing ultraviolet–visible spectra of these two isomers, as higher Clar pi-sextets are associated with larger HOMO-LUMO gaps; the highest-wavelength absorbance of phenanthrene is at 293 nm, while anthracene is at 374 nm. Three Clar structures with two sextets each are present in the four-ring chrysene structure: one having sextets in the first and third rings, one in the second and fourth rings, and one in the first and fourth rings. Superposition of these structures reveals that the aromaticity in the outer rings is greater (each has a sextet in two of the three Clar structures) compared to the inner rings (each has a sextet in only one of the three).

Polycyclic aromatic compounds characteristically reduce to the radical anions. The redox potential correlates with the size of the PAH.

Table: Half-cell potential of aromatic compounds against the SCE.

Compound	Potential (V)
benzene	−3.42
biphenyl	−2.60
naphthalene	−2.51
anthracene	−1.96
phenanthrene	−2.46
perylene	−1.67
pentacene	−1.35

Sources and Distribution

Polycyclic aromatic hydrocarbons are primarily found in natural sources such as creosote. They can result from the incomplete combustion of organic matter. PAHs can also be produced geologically when organic sediments are chemically transformed into fossil fuels such as oil and coal. PAHs are considered ubiquitous in the environment and can be formed from either natural or manmade combustion sources. The dominant sources of PAHs in the environment are thus from human activity: wood-burning and combustion of other biofuels such as dung or crop residues contribute more than half of annual global PAH emissions, particularly due to biofuel use in India and China. As of 2004, industrial processes and the extraction and use of fossil fuels made up slightly more than one quarter of global PAH emissions, dominating outputs in industrial countries such as the United States. Wildfires are another notable source. Substantially higher outdoor air, soil, and water concentrations of PAHs have been measured in Asia, Africa, and Latin America than in Europe, Australia, the U.S., and Canada.

PAHs are typically found as complex mixtures. Lower-temperature combustion, such as tobacco smoking or wood-burning, tends to generate low molecular weight PAHs, whereas high-temperature industrial processes typically generate PAHs with higher molecular weights.

In the Aqueous Environment

Most PAHs are insoluble in water, which limits their mobility in the environment, although PAHs sorb to fine-grained organic-rich sediments. Aqueous solubility of PAHs decreases approximately logarithmically as molecular mass increases. Two-ringed PAHs, and to a lesser extent three-ringed PAHs, dissolve in water, making them more available for biological uptake and degradation. Further, two- to four-ringed PAHs volatilize sufficiently to appear in the atmosphere predominantly in gaseous form, although the physical state of four-ring PAHs can depend on temperature. In contrast, compounds with five or more rings have low solubility in water and low volatility; they are therefore predominantly in solid state, bound to particulate air pollution, soils, or sediments. In solid state, these compounds are less accessible for biological uptake or degradation, increasing their persistence in the environment.

Human Exposure

A wood-burning open-air cooking stove. Smoke from solid fuels like wood is a large source of PAHs globally.

Human exposure varies across the globe and depends on factors such as smoking rates, fuel types in cooking, and pollution controls on power plants, industrial processes, and vehicles. Developed countries with stricter air and water pollution controls, cleaner sources of cooking (i.e., gas and electricity vs. coal or biofuels), and prohibitions of public smoking tend to have lower levels of PAH exposure, while developing and undeveloped countries tend to have higher levels. Surgical smoke plume have been proven to contain PAHs in several independent research studies. For indoor contaminants, surgical plume needs to be noticed as a serious potential health risk for the 59 million health care workers around the world.

Burning solid fuels such as coal and biofuels in the home for cooking and heating is a dominant global source of PAH emissions that in developing countries leads to high levels of exposure to indoor particulate air pollution containing PAHs, particularly for women and children who spend more time in the home or cooking.

In industrial countries, people who smoke tobacco products, or who are exposed to second-hand smoke, are among the most highly exposed groups; tobacco smoke contributes to 90% of indoor PAH levels in the homes of smokers. For the general population in developed countries, the diet is otherwise the dominant source of PAH exposure, particularly from smoking or grilling meat or consuming PAHs deposited on plant foods, especially broad-leafed vegetables, during growth. PAHs are typically at low concentrations in drinking water.

Smog in Cairo. Particulate air pollution, including smog, is a substantial cause of human exposure to PAHs.

Emissions from vehicles such as cars and trucks can be a substantial outdoor source of PAHs in particulate air pollution. Geographically, major roadways are thus sources of PAHs, which may distribute in the atmosphere or deposit nearby. Catalytic converters are estimated to reduce PAH emissions from gasoline-fired vehicles by 25-fold.

People can also be occupationally exposed during work that involves fossil fuels or their derivatives, wood-burning, carbon electrodes, or exposure to diesel exhaust. Industrial activity that can produce and distribute PAHs includes aluminum, iron, and steel manufacturing; coal gasification, tar distillation, shale oil extraction; production of coke, creosote, carbon black, and calcium carbide; road paving and asphalt manufacturing; rubber tire production; manufacturing or use of metal working fluids; and activity of coal or natural gas power stations.

Environmental Distribution and Degradation

Crude oil on a beach after a oil spill.

PAHs typically disperse from urban and suburban non-point sources through road run-off, sewage, and atmospheric circulation and subsequent deposition of particulate air pollution. Soil and river sediment near industrial sites such as creosote manufacturing facilities can be highly contaminated with PAHs. Oil spills, creosote, coal mining dust, and other fossil fuel sources can also distribute PAHs in the environment.

Two- and three-ringed PAHs can disperse widely while dissolved in water or as gases in the atmosphere, while PAHs with higher molecular weights can disperse locally or regionally adhered to particulate matter that is suspended in air or water until the particles land or settle out of the water column. PAHs have a strong affinity for organic carbon, and thus highly organic sediments in rivers, lakes, and the ocean can be a substantial sink for PAHs.

Algae and some invertebrates such as protozoans, mollusks, and many polychaetes have limited ability to metabolize PAHs and bioaccumulate disproportionate concentrations of PAHs in their tissues; however, PAH metabolism can vary substantially across invertebrate species. Most vertebrates metabolize and excrete PAHs relatively rapidly. Tissue concentrations of PAHs do not increase (biomagnify) from the lowest to highest levels of food chains.

PAHs transform slowly to a wide range of degradation products. Biological degradation by microbes is a dominant form of PAH transformation in the environment. Soil-consuming invertebrates such as earthworms speed PAH degradation, either through direct metabolism or by improving the conditions for microbial transformations. Abiotic degradation in the atmosphere and the top layers of surface waters can produce nitrogenated, halogenated, hydroxylated, and oxygenated PAHs; some of these compounds can be more toxic, water-soluble, and mobile than their parent PAHs.

PAHs in Urban Soils

The British Geological Survey reported the amount and distribution of PAH compounds including parent and alkylated forms in urban soils at 76 locations in Greater

London. The study showed that parent (16 PAH) content ranged from 4 to 67 mg/kg (dry soil weight) and an average PAH concentration of 18 mg/kg (dry soil weight) whereas the total PAH content (33 PAH) ranged from 6 to 88 mg/kg and fluoranthene and pyrene were generally the most abundant PAHs. Benzo[*a*]pyrene (B*a*P), the most toxic of the parent PAHs, is widely considered a key marker PAH for environmental assessments; the normal background concentration of B*a*P in the London urban sites was 6.9 mg/kg (dry soil weight). London soils contained more stable four- to six-ringed PAHs which were indicative of combustion and pyrolytic sources, such as coal and oil burning and traffic-sourced particulates. However, the overall distribution also suggested that the PAHs in London soils had undergone weathering and been modified by a variety of pre-and post-depositional processes such as volatilization and microbial biodegradation.

PAHs in Peatlands

Managed burning of moorland vegetation in the UK has been shown to generate PAHs which become incorporated into the peat surface. Burning of moorland vegetation such as heather initially generates high amounts of two- and three-ringed PAHs relative to four- to six-ringed PAHs in surface sediments, however, this pattern is reversed as the lower molecular weight PAHs are attenuated by biotic decay and photodegradation. Evaluation of the PAH distributions using statistical methods such as principal component analyses (PCA) enabled the study to link the source (burnt moorland) to pathway (suspended stream sediment) to the depositional sink (reservoir bed).

PAHs in Rivers, Estuarine and Coastal Sediments

Concentrations of PAHs in river and estuarine sediments vary according to a variety of factors including proximity to municipal and industrial discharge points, wind direction and distance from major urban roadways, as well as tidal regime which controls the diluting effect of generally cleaner marine sediments relative to freshwater discharge. Consequently, the concentrations of pollutants in estuaries tends to decrease at the river mouth. Understanding of sediment hosted PAHs in estuaries is important for the protection of commercial fisheries (such as mussels) and general environmental habitat conservation because PAHs can impact the health of suspension and sediment feeding organism. River-estuary surface sediments in the UK tend to have a lower PAH content than sediments buried 10–60 cm from the surface reflecting lower present day industrial activity combined with improvement in environmental legislation of PAH. Typical PAH concentrations in UK estuaries range from about 19 to 16,163 µg/kg (dry sediment weight) in the River Clyde and 626 to 3,766 µg/kg in the River Mersey. In general estuarine sediments with a higher natural total organic carbon content (TOC) tend to accumulate PAHs due to high sorption capacity of organic matter. A similar correspondence between PAHs and TOC has also been observed in the sediments of tropical mangroves located on the coast of southern China.

Minor Sources

Volcanic eruptions may emit PAHs. Certain PAHs such as perylene can also be generated in anaerobic sediments from existing organic material, although it remains undetermined whether abiotic or microbial processes drive their production.

Human Health

Cancer is a primary human health risk of exposure to PAHs. Exposure to PAHs has also been linked with cardiovascular disease and poor fetal development.

Cancer

PAHs have been linked to skin, lung, bladder, liver, and stomach cancers in well-established animal model studies.

PAHs.

Historically, PAHs contributed substantially to our understanding of adverse health effects from exposures to environmental contaminants, including chemical carcinogenesis. In 1775, Percivall Pott, a surgeon at St. Bartholomew's Hospital in London, observed that scrotal cancer was unusually common in chimney sweepers and proposed the cause as occupational exposure to soot. A century later, Richard von Volkmann reported increased skin cancers in workers of the coal tar industry of Germany, and by the early 1900s increased rates of cancer from exposure to soot and coal tar was widely accepted. In 1915, Yamigawa and Ichicawa were the first to experimentally produce cancers, specifically of the skin, by topically applying coal tar to rabbit ears.

In 1922, Ernest Kennaway determined that the carcinogenic component of coal tar mixtures was an organic compound consisting of only carbon and hydrogen. This component was later linked to a characteristic fluorescent pattern that was similar but not identical to benz[*a*]anthracene, a PAH that was subsequently demonstrated to cause tumors. Cook, Hewett and Hieger then linked the specific spectroscopic fluorescent profile of benzo[*a*]pyrene to that of the carcinogenic component of coal tar, the first time that a specific compound from an environmental mixture (coal tar) was demonstrated to be carcinogenic.

In the 1930s and later, epidemiologists from Japan, the UK, and the US, including Richard Doll and various others, reported greater rates of death from lung cancer following occupational exposure to PAH-rich environments among workers in coke ovens and coal carbonization and gasification processes.

Mechanisms of Carcinogenesis

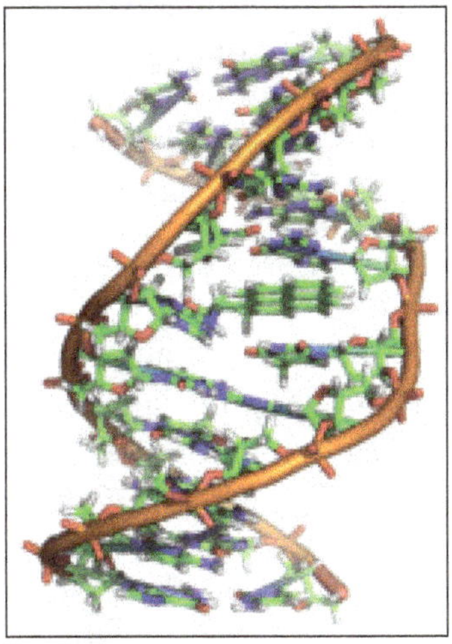

An adduct formed between a DNA strand and an epoxide derived from a benzo[*a*]pyrene molecule (center); such adducts may interfere with normal DNA replication.

The structure of a PAH influences whether and how the individual compound is carcinogenic. Some carcinogenic PAHs are genotoxic and induce mutations that initiate cancer; others are not genotoxic and instead affect cancer promotion or progression.

PAHs that affect cancer initiation are typically first chemically modified by enzymes into metabolites that react with DNA, leading to mutations. When the DNA sequence is altered in genes that regulate cell replication, cancer can result. Mutagenic PAHs, such as benzo[*a*]pyrene, usually have four or more aromatic rings as well as a "bay region", a structural pocket that increases reactivity of the molecule to the metabolizing enzymes. Mutagenic metabolites of PAHs include diol epoxides, quinones, and radical PAH cations. These metabolites can bind to DNA at specific sites, forming bulky complexes called DNA adducts that can be stable or unstable. Stable adducts may lead to DNA replication errors, while unstable adducts react with the DNA strand, removing a purine base (either adenine or guanine). Such mutations, if they are not repaired, can transform genes encoding for normal cell signaling proteins into cancer-causing oncogenes. Quinones can also repeatedly generate reactive oxygen species that may independently damage DNA.

Enzymes in the cytochrome family (CYP1A1, CYP1A2, CYP1B1) metabolize PAHs to diol epoxides. PAH exposure can increase production of the cytochrome enzymes, allowing the enzymes to convert PAHs into mutagenic diol epoxides at greater rates. In this pathway, PAH molecules bind to the aryl hydrocarbon receptor (AhR) and activate it as a transcription factor that increases production of the cytochrome enzymes. The activity of these enzymes may at times conversely protect against PAH toxicity, which is not yet well understood.

Low molecular weight PAHs, with two to four aromatic hydrocarbon rings, are more potent as co-carcinogens during the promotional stage of cancer. In this stage, an initiated cell (a cell that has retained a carcinogenic mutation in a key gene related to cell replication) is removed from growth-suppressing signals from its neighboring cells and begins to clonally replicate. Low-molecular-weight PAHs that have bay or bay-like regions can dysregulate gap junction channels, interfering with intercellular communication, and also affect mitogen-activated protein kinases that activate transcription factors involved in cell proliferation. Closure of gap junction protein channels is a normal precursor to cell division. Excessive closure of these channels after exposure to PAHs results in removing a cell from the normal growth-regulating signals imposed by its local community of cells, thus allowing initiated cancerous cells to replicate. These PAHs do not need to be enzymatically metabolized first. Low molecular weight PAHs are prevalent in the environment, thus posing a significant risk to human health at the promotional phases of cancer.

Cardiovascular Disease

Adult exposure to PAHs has been linked to cardiovascular disease. PAHs are among the complex suite of contaminants in tobacco smoke and particulate air pollution and may contribute to cardiovascular disease resulting from such exposures.

In laboratory experiments, animals exposed to certain PAHs have shown increased development of plaques (atherogenesis) within arteries. Potential mechanisms for the pathogenesis and development of atherosclerotic plaques may be similar to the mechanisms involved in the carcinogenic and mutagenic properties of PAHs. A leading hypothesis is that PAHs may activate the cytochrome enzyme CYP1B1 in vascular smooth muscle cells. This enzyme then metabolically processes the PAHs to quinone metabolites that bind to DNA in reactive adducts that remove purine bases. The resulting mutations may contribute to unregulated growth of vascular smooth muscle cells or to their migration to the inside of the artery, which are steps in plaque formation. These quinone metabolites also generate reactive oxygen species that may alter the activity of genes that affect plaque formation.

Oxidative stress following PAH exposure could also result in cardiovascular disease by causing inflammation, which has been recognized as an important factor in the development of atherosclerosis and cardiovascular disease. Biomarkers of exposure to PAHs in humans have been associated with inflammatory biomarkers that are recognized as important predictors of cardiovascular disease, suggesting that oxidative stress resulting from exposure to PAHs may be a mechanism of cardiovascular disease in humans.

Developmental Impacts

Multiple epidemiological studies of people living in Europe, the United States, and China have linked in utero exposure to PAHs, through air pollution or parental occupational exposure, with poor fetal growth, reduced immune function, and poorer neurological development, including lower IQ.

ORGANOCHLORIDE

Organochlorines (OC) are a group of chlorinated compounds widely used as pesticides. These chemicals belong to the class of persistent organic pollutants (POPs) with high persistence in the environment. OC insecticides were earlier successfully used in control of malaria and typhus, yet they are banned in most of the advanced countries. The review statistics on the use of different pesticides shows that 40% of all pesticides used belong to the organochlorine class of chemicals. Due to their low cost and the need against various pests, organochlorine insecticides such as DDT, hexachlorocyclohexane (HCH), aldrin and dieldrin are among the most widely used pesticides in developing countries of Asia.

Organophosphates

Organophosphates (OP) are esters of phosphoric acid. The OP group of pesticides asserts its effects through irreversible inactivation of the enzyme acteylcholinesterase, which is essential for nerve function in humans, insects and many other animals. OP samples degrade rapidly by hydrolysis on exposure to light, air and soil, however small amounts are detected in food and drinking water.

Carbamates

Carbamates are organic compounds derived from carbamic acid (NH_2COOH). The functional group present in carbamate insecticides are carbamate esters. Their mechanism of action is by reversible inactivation of the enzyme acteylcholinesterase. Carbamates break down in the environment within weeks or months.

Pyrethroides

Pyrethroides and pyrethrins are similar organic compounds isolated from the flowers of pyrethrums (Chrysanthemum Coccineum and C. cinerariaefolium). The insecticidal properties of pyrethrins are derived from ketoalcoholic esters of chrysanthemic and pyrethroic acids. Pyrethroides affect the sodium channels and lead to paralysis of the organism. Pyrethroides have a comparatively slight level of mammalian toxicity and have a fast biodegradation capacity. Exposure to very high levels of the compounds in air, food or water may cause giddiness, headache, vomiting, muscle twitching, low energy, convulsions and loss of consciousness.

Phenylamides

Phenylamide fungicides are systemic compounds that show potent eradicative anti-fungal activity. When added to the soil, they enhance plant growth and yield; in addition, these fungicides affect the homeostastis of the soil system. These chemicals affect nutrient cycling and enter the food chain, and have thus been reported to affect higher organisms including humans. The affect nucleic acids by inhibiting the activity of RNA polymerase I system. They are known to impact mitosis and cell division in target fungi.

Phenoxyalkonates

Phenoxyalkonates are a widely used family of herbicides. These pesticides are mainly used to control weeds in agriculture. Nearly all compounds of this group are degraded by microorganisms.

Triazines

The compounds that fall under this category are herbicidal pesticides. They include desmetryne, chlorazine, atriazine, propazine, etc. These compounds are known to have potential use as insect chemosterilants. Higher concentrations of these herbicides were found to inhibit plant catabolism pathway.

Benzoic Acid

Benzoic acid herbicides include dicamba, dichlobenil, chlorambin, bromoxynil, ioxynil and naptalam. Little information is available regarding their degradation by soil microbes. Ioxynil is found to precipitate in acid soils.

Phthalimide

Phthalimides include three fungicides, captan, folpet and captafol which together represent the second most important group of organic fungicides used in American agriculture. They represent about half the usage of the dithiocarbamates. The fungicides difolatan, captan and folpet react with thiols such as cysteine and glutathione at acidic pH levels of 4.0 to 5.0.

Dipyrids

The dipyridyl herbicides include paraquat and diquat. They are strongly adsorbed as organic cations in the soil. Microorganisms metabolize paraquat as the main source of nitrogen.

Others

There are many more pesticides used in agricultural practice. Heavy metals have found vast use as pesticides. Elements like iron, lead, sulphur, arsenic, mercury, zinc, tin, etc. have been used in inorganic or organic metal form. Methyl mercuric chloride, sodium arsenate, calcium arsenate, zinc phosphide are some of the compounds that fall under this category. Among the various classes of pesticides, organochlorines and organophosphates are widely used. Organochlorines are known for their high persistence and toxicity characteristics. These pesticides cause neurological damage, endocrine disorders, and have acute and chronic health effects. Hence contamination of the environment with organochlorine pesticides drastically affects the ecosystem.

Organochlorine Pesticides – Chemistry, Persistence and Hazard Classification

The basic characteristics of organochlorine pesticides are high persistence, low polarity, low aqueous solubility and high lipid solubility. Organochlorine pesticides can enter the environment after pesticide applications, polluted wastes discarded into landfills, and discharges from industrial units that synthesize these chemicals. They are volatile and stable; some can adhere to the soil and air, thus increasing the chances of high persistence in the environment, and are identified as agents of chronic exposure to animals and humans.

They have a related chemical structure, showing chlorine substituted aliphatic or aromatic rings. Due to their structural resemblances, these compounds share certain physicochemical characteristics such as persistence, bioaccumulation and toxicity. One basic character that they share across the spectrum is persistence, where persistence is defined as half-life greater than two months in water or six months in soil sediment. The persistence of OC compounds varies from moderate persistence with half-life of approximately 60 days to high persistence with half-life up to 10–15 years. The most commonly used pesticide in agricultural practice is dichlorodiphenyltrichloroethane (DDT), which is moderately hazardous, with high persistence and a half-life of 2–15 years. The use of DDT is now banned in many countries but it is illegally used in most of the developing countries. This applies also to endosulphan, an insecticide which is highly hazardous and has moderate persistence with a half-life of fifty days and is used in the production of cashew.

Due to the high persistence and bioaccumulation potential, the Stockholm Convention has classified most of the OC compounds as environmental hazards and banned the use of many of them. However in many developing countries they are still in use making the ban ineffective.

Biochemical Toxicity of Organochlorines

Organochlorine toxicity is mainly due to stimulation of the central nervous system. Cyclodines, such as the GABA antagonists endosulphan and lindane, inhibit the calcium ion influx and Ca- and Mg-ATPase causing release of neurotransmittors. Epidemiological studies have exposed the etiological relationship between Parkinson's disease and organochlorine pollutants.

Effect in Humans

Examination of effects of different classes of pesticides leads to the conclusion that many of them are responsible for hypertension, cardiovascular disorders and other health related problems in humans. Organochlorines act as endocrine disrupting chemicals (EDCs) by interfering with molecular circuitry and function of the endocrine system. Farm workers, their families and those who pass through a region applied with pesticides can absorb a measurable quantity of pesticides. The presence of pesticide residues has been detected in blood plasma of workers in agricultural farms. Direct or indirect exposure to pesticides leads to neuromuscular disorders and stimulation of drug and steroid metabolism.

Another mode of exposure to these pesticides is through diet. Among food items, fatty food such as meat, fish, poultry, and dairy products serve as main causes. Many of the organochlorine molecules are carcinogens and neurotoxic. The hazardous nature of organochlorines was explained by citing different examples. The menace caused by endosulfan is of great concern. Endosulfan remains in the environment for longer periods and bio-accumulates in plants and animals which leads to contamination of food consumed by humans. It affects mainly the central nervous system and was found to have higher acute inhalation toxicity than dermal toxicity. Gastrointestinal absorption of endosulfan is very high.

Disproportion of thyroid hormones can lead to a variety of disorders. Serum concentrations of p-p'-DDE and HCB were found to be associated with abnormal thyroid hormone levels. p,p'-DDE was reported to increase free thyroxine (T4) and total triiodothyronine (T3) levels, and to be inversely associated with thyroid-stimulating hormone (TSH). On exposure to dioxinlike organochlorines, a dose-dependent decrease in total T4 was also reported. Organochlorine pesticides were reported to increase the risk of hormone-related cancers including breast, prostate, stomach and lung cancer. Recently dioxins have been found in human ovarian follicular fluid, which may lead to the development of endometriosis. Exposure to dioxins can cause several autoimmune diseases, including multiple sclerosis and eczema. Organochlorines can function as xenoestrogens and compounds such as TCDD, methoxychlor and alachlor were reported to exert effects on human and experimental animals due to inhibited synthesis and increased degradation of thyroid hormones.

Analysis of the National Health and Nutrition Examination Survey 1999–2004 studying the relation between organochlorine pesticides and prostate and breast cancers has shown that serum concentrations of b-HCH, trans-nonachlor, and dieldrin were significantly associated with prostate cancer prevalence. In children, exposure to dioxins showed significant positive associations with learning disability (LD). Risk of attention deficit hyperactivity disorder (ADHD) at higher levels of p,p'-DDE and PCBs exposure was reported. Prenatal exposure to p,p'-DDE and its presence in cord serum was found to lead to disappearance of neuronal development after 12 months of infant age. Epidemiological studies have shown that exposure to persistent organic pollutants, mainly organochlorine pesticides, is strongly associated with type 2 diabetes. Some persistent organic pollutants, as highly chlorinated PCBs and trans-nonachlor, were associated with the incidence of type 2 diabetes in obese people.

Selected persistent organic pollutants are reported to induce divergent actions on blood pressure, suggesting a chemical structure based association of pesticides. In a population based study, different persistent organic pollutants and pesticides were reported to be associated with liver dysfunction biomarkers such as bilirubin, ALT and ALP, suggesting that these environmental pollutants can cause adverse effects on liver functions. A study conducted in Costa Rica reported that occupational pesticide exposure to dialdrin could be partly responsible for the increased risk of Parkinson's disease seen in the population. Studies showed that the change of lipids over time, especially LDL-cholesterol, is linked to POP exposure. Increased oxidative stress markers in plasma were found to be associated with exposure of POPs and could be a causative agent for oxidative stress. Persistent organic pollutants were reported to influence the complement system, leading to activation of the immune system in humans. Detection of organochlorine pesticides from human breast milk was reported from many places in the world. In Croatia, p,p'-DDE was found to be the dominant organochlorine pesticide in human breast milk. Exposure of infants to chlordanes via breast milk was reported as a potential health risk in Korea. Another study from Korea also revealed the presence of organochlorine pesticides (OCPs) chlordanes, aldrin, dichlorodiphenyltrichloroethanes (DDTs), dieldrin, heptachlors, endrins, hexachlorocyclohexanes (HCHs), hexachlorobenzene (HCB), toxaphenes and mirex, in milk. Organochlorine pesticides HCB, β-HCH, pp'DDE, pp'DDT, pp'DDT, Σ-DDT were present in breast milk of the population in Guerrero, Mexico, proportionally to exposure.

A study conducted in China showed that prenatal exposure to DDT, β-BHC, HCB and mirex caused decrease in birth weight of infants. A number of studies were published on the effect of organochlorine pesticides on induction of diabetes mellitus in humans. A recent study reported that POP exposure is a risk factor contributing to insulin resistance. Chronic exposure to chlordecone was found to cause hypertensive disorders in pregnancy and gestational diabetes mellitus among French Caribbean women. In a study conducted in Slovakia, highly increased blood levels of diabetes (fasting glucose and insulin) and obesity markers (BMI, triglyceride and cholesterol) were found in large groups of males and females in highly polluted areas. A significant decrease in testosterone level was also observed in males. Prevalence of type 2 diabetes and exposure to persistent organic pollutants has been established. Recent studies on organochlorine pesticides have shown that β-HCH, HCB and DDT residues bio-accumulate in maternal and cord sera and from maternal blood they can be transferred through the placenta and affect thyroid hormone levels in the newborn. OC pesticides have been suggested to affect the thyroid system through gender-specific mechanisms; the extent of the effect may differ among compounds. A report from Brazil had shown that OC compounds are reported to trigger anti-androgenic effects in men and estrogenic effects in women. OC pesticide heptachlor was reported to induce mitochondria-mediated cell death via impairing electron transport chain complex III, thus acting as a neurotoxicant with possible association with Parkinson's disease.

Exposure to organochlorine pesticide residues was reported as a potential risk factor for gallstone disease in humans. Potential neurotoxic effects of organochlorine compounds were reported on early psychomotor development even at low doses. A positive correlation was observed of exposure to some OC pesticides and vitamin D deficiency in humans. Early exposure to certain environmental chemicals, especially organochlorine compounds, with endocrine-disruption activity were reported to interfere with neonatal thyroid hormone status.

Toxic Effect of Pesticides in Fauna

Wild birds are of great importance to the ecosystem. Decline in the bird community serves as an indicator of environmental pollution. Continuous use of pesticides is one of the major causes for the reduction of birds. In many cases the impact is not direct, however repetitive use of pesticides like DDT in soil is taken up by earthworms which are then ingested by birds and thus their accumulation may result in a large loss in bird population. Subsequent research has also identified other pesticides and industrial chemicals that cause mortality and reproductive impairment, which affects both embryos and adult birds. The effects on embryos include mortality or reduced hatchability, wasting syndrome and teratological effects that produce skeletal abnormalities and impaired differentiation of the reproductive and nervous systems through mechanisms of hormonal mimicking of estrogens. The range of chemical effects on adult birds covers acute mortality, sub-lethal stress, reduced fertility, suppression of egg formation, eggshell thinning and impaired incubation and chick rearing behaviors. Pesticides cause extinction, behavioral changes, loss of safe habitat and population decline in several birds. Prolonged use of pesticides causes a drastic decrease in birds like the peregrine falcon, sparrow hawk and bald eagle. The levels of organochlorines in seabird eggs were indicated by forming a deposit of pollutants in the body, thus serving as a useful indicator of environmental contamination.

Toxic Effect in Farm Animals

The prolonged use of pesticides in agriculture has caused serious health problems as these pesticides accumulate and affect the food chain. Organochlorine compounds are highly lipophilic and can accumulate in fat-rich food such as meat and milk. Pesticides are introduced into cattle mainly through fodder or contaminated water used for household and public purposes.

Amphibians and insectivorous reptiles, like lizards, have an important function in linking invertebrates with vertebrates in the food chain. They serve as a food source for some organisms and are also a means by which chemical residues, especially residues of organochlorine pesticides taken in with contaminated prey, can enter food chains. Amphibians consume these pesticides by a number of ways, including inhalation, contact and through ingestion. Amphibians in open water bodies may also be exposed to pesticides due to run-off from adjacent agricultural land on which chemicals are used to control crop pests. Continuous exposure of honey bees to pesticides affects the quality of honey. The routes of honey contamination with pesticides are direct and indirect. The direct is treatment of beehives with pesticides. Wild animals, including the grasscutter (Thryonomys swinderianus), which are a good source of protein, are seriously affected by the use of pesticides. Grasscutters are a source of food for the people of Ghana in Africa. As pesticides have high effect on the animal and bird community, ultimately humans also take up pesticides as meat, milk and crops derived from these animals and plants are consumed by humans.

References

- Dioxins-and-their-effects-on-human-health, fact-sheets, news-room: who.int, Retrieved 09 March, 2020
- Harley, K.; Marks, A.; Chevrier, J.; Bradman, A.; Sjödin, A.; Eskenazi, B. (2010). "PBDE Concentrations in Women's Serum and Fecundability". Environmental Health Perspectives. 118 (5): 699–704. doi:10.1289/ehp.0901450. PMC 2866688. PMID 20103495
- Fetzer, John C. (16 April 2007). "The chemistry and analysis of large PAHs". Polycyclic Aromatic Compounds. 27 (2): 143–162. doi:10.1080/10406630701268255
- Weil, ED; Levchik, SV (2015). Flame Retardants for Plastics and Textiles: Practical Applications. Munich: Carl Hanser Verlag. p. 97. ISBN 978-1569905784
- Alaluusua S, Calderara P, Gerthoux PM, et al. (2004). "Developmental dental aberrations after the dioxin accident in Seveso". Environ. Health Perspect. 112 (13): 1313–8. doi:10.1289/ehp.6920. PMC 1247522. PMID 15345345

Bioaccumulation and Biomagnification

CHAPTER 3

The accumulation of pesticides and other chemicals in an organism is termed as bioaccumulation. Biomagnification refers to the high levels of concentration of toxins and chemicals in the tissue of the tolerant organism. This chapter closely examines bioaccumulation and biomagnification to provide an extensive understanding of the subject.

BIOACCUMULATION

Bioaccumulation describes the accumulation and enrichment of contaminants in organisms, relative to that in the environment. Bioaccumulation is the net result of all uptake and loss processes, such as respiratory and dietary uptake, and loss by egestion, passive diffusion, metabolism, transfer to offspring and growth. Bioaccumulation thereby comprises the more specific processes of bioconcentration and biomagnification. Bioconcentration is the process of direct partitioning of chemicals between the water and the organism, leading to elevated concentrations in the latter. Biomagnification is the result of contaminant uptake from the diet leading to higher concentrations in the feeder than in the diet. Subsequently, biomagnification leads to increased chemical concentration with higher trophic position in the food web.

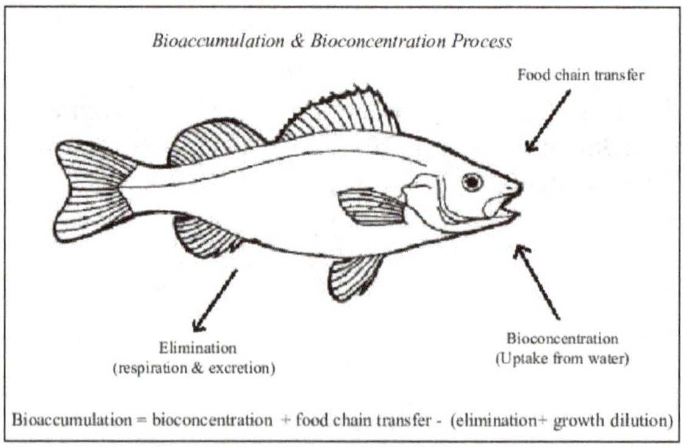

Bioconcentration and Bioaccumulation.

Bioaccumulation of contaminants (dots) to an organism (fish) as a net result of uptake and loss processes (arrows). Uptake is direct from the water by respiration and indirect through the diet. Examples of loss processes are respiration, metabolism, egestion, and growth dilution. Bioaccumulating chemicals increase more than 5000 times from the water to the animal. As the total biomass decreases per trophic level in the food chain (while the contaminants remain), the contaminant concentrations increase moving up the food chain.

Bioaccumulation is defined as the increase of contaminant concentrations in aquatic organisms following uptake from the ambient environmental medium. Different sources of exposure contribute to contaminant bioaccumulation. Toxicokinetic parameters are critical in the bioaccumulation processes, including the dissolved uptake rate constant, dietary assimilation efficiency, and efflux. Much study has been conducted on the quantification of these biokinetic parameters in diverse marine organisms. A well-developed biokinetic model then incorporates these biokinetic processes as well as contaminant geochemistry and animal physiology to predict bioaccumulation of contaminants in aquatic organisms. There are many applications of the biokinetic model in aquatic ecotoxicology, especially in understanding the change of contaminant concentration, which forms a fundamental basis in aquatic toxicology. Although biomonitoring programs have been established for decades, there are several caveats underlying the application of biomonitors given the multiple ecosystem stresses currently faced by coastal and estuarine ecosystems.

Bioaccumulation quantification describes how it can be that concentrations in biota sometimes are higher than those in the surrounding environment or in the prey or food of organisms. The general model for bioaccumulation of hydrophobic organic chemicals stipulates that organic chemicals bioconcentrate in lipid tissues mainly by passive exchange processes from and to water. The magnitude of bioconcentration largely depends on the hydrophobicity, expressed via the octanol–water partition coefficient, and the lipid content of the organisms. Laboratory and field studies indicate that phthalate esters do not biomagnify in aquatic food webs:

- High molecular weight phthalates show evidence of trophic dilution in aquatic food webs. Laboratory and modeling studies indicate that metabolic transformation is a key mitigating factor in this respect.
- Phthalate esters with intermediate molecular weights (like DBP, diisobutyl phthalate, and butyl benzyl phthalate) have bioaccumulation patterns that are consistent with the general lipid–water partitioning model.
- Low molecular weight phthalates display bioaccumulation factors that are greater than predicted from a lipid–water partitioning model, and a considerable variation in field-derived bioaccumulation factors for low molecular weight phthalate esters across aquatic species suggests that species-specific differences in metabolic transformation capacity have a significant effect on observed bioaccumulation.

With some exceptions, bioaccumulation factors of phthalates are below 5000. The low bioavailability of the high molecular weight phthalate esters, as a consequence of their low water solubility and strong adsorption to (dissolved) organic carbon, is the main reason why these compounds too do not bioaccumulate to a considerable extent. It should be noted that although theoretical studies show that as much as 60% of the exposure of high molecular weight phthalates in predators could be derived from the diet, only limited data on dietary bioaccumulation have been reported.

Some organic pesticides are the most widespread POPs (Persistent Organic Pollutants) used in agriculture worldwide. Other POPs such as volatilized industrial pollutants as well as pollutants from fuel combustion can also contaminate large areas due to dispersion by wind and rainfall. Non-biodegradable molecules persist in the environment and the soil serves as a large sink for them. These pollutants are able to move in the environment and in low concentrations over a long period of time, but through bioaccumulation they can have hazardous effects on the biota of the region.

Adsorption or not of a molecule in the soil is a key step to define bioavailability. The concept of "preferential partition" helps us understand the competitive mechanisms between the soil and biosphere. Bioaccumulation occur when concentrations in the biosphere are much higher than those found in the surrounding environment. The uptake of persistent molecules to levels of high concentrations can have hazardous effects on flora and, through the food chain, on fauna and human health.

Xenobiotics not adsorbed by the soil can become mobile and when the molecule is persistent it may be dispersed by water or air through the biosphere, polluting large geographical areas and affecting the biodiversity of the flora and fauna. The application of pesticides over large areas can only be sustainable if their molecules biodegrade naturally. Remediation can be applied usefully to accidents in restricted areas but it is not feasible over large areas due to the high costs. The only efficient policy for recalcitrant molecules that cannot be controlled by remediation is the restriction of their use or banning them altogether.

How Molecules can Escape Degradation in the Environment

The adsorption of molecules, due to their affinity to other chemical components of the soil matrix or soil biosphere, protects them from chemical degradation as well biodegradation. Non adsorbed molecules are bioavailable and therefore are exposed to degradation. Experiments with soils weathered for long time with ethylene bromide contamination in which was added ^{14}C-ethylene bromide, shows strong biodegradation of the new applied radioactive substance after sixty days at the same time that weathered contamination substance remains in the same concentration. Microorganisms involved in biodegradation are active against a freshly applied radioactive molecule but are not efficient for the same substance entrapped in the soil. Weathered molecules by soil friction process in which the larger soil particles are broken into smaller ones results in the gradually desorption of the molecules with enhanced bioavalability. Therefore it is a pitfall to conclude that the hazardous effects are only caused by pesticide concentrations in the soil.

The fate of pollutants in the soil depends on the soil properties and the physical chemical characteristics of these molecules. Essentially the uptake of pollutants in the soil can be understood to be the result of chemical attraction and bond strength. These processes in the soil are governed by a number of well described phenomena including: Van der Waals forces, hydrogen bonding, ion exchange, charge transfer mechanisms, lipophilic affinity, entrapment and covalent reactions to humic acids. In essence, pollutant molecules will move to the most attractive sites in the soil environment. Competition between plant roots, soil mesofauna, microbes and soil organic and inorganic components makes residue to bind in the most attractive site where residues will accumulate preferentially. Pollutants, such as pesticides, which are not held or bound to the living and non-living parts of the soil, will be leached through the soil, resulting in pollution of ground water, rivers and reservoirs or they will be volatilized.

Where Molecules Adsorb

In aquatic environments, where polar water molecules predominate, lipophilic pollutants have affinity for organic matter and therefore always move toward the biosphere. In this case the bioaccumulation factor depends on molecular size and correlates positively with a lipophilic character measured by the K_{ow} value. However, polarity on its own cannot explain to which compartment POP's move to in the soil.

Soils are a highly variable mixture of mineral and organic materials with living, dead and decaying biologic components. There is a lipophilic fraction as well that can adsorb lipophilic pollutants. The binding process is complex considering the diversity of compartments such as microorganisms, mesofauna and plants that compete for the uptake of lipophilic pollutants. Despite difficulties to standardize a methodology for lipid determination, this parameter is used to calculate the bioaccumulation factor. Soils can function as a filter when they adsorb the remaining residues, which then become unavailable for the biosphere. On the other hand low adsorption capacity pollutants remain bioavailable and can contaminate water, air, fodder plants, livestock and moving along the food chain up to humans. The soils may function as a filter or as a source of pollutants and this depends mainly on the kind of soil.

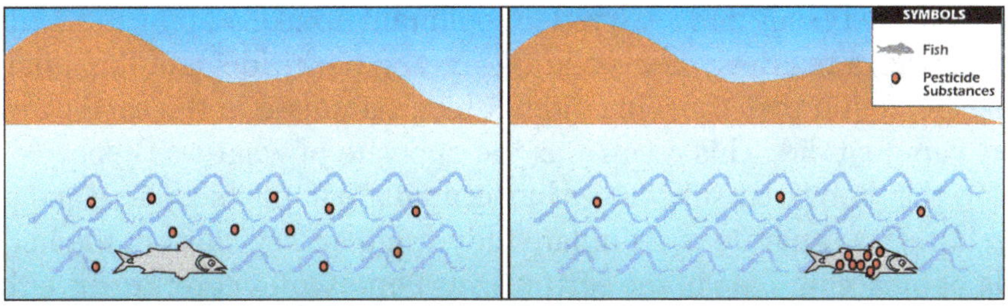

In aquatic systems the movement of lipophilic substance is unidirectional towards living forms, resulting in bioaccumulation.

Despite enormous amounts of published scientific literature about bioaccumulation, the distribution of pollutants in the soil or biosphere is not well understood.

Lipophilic Molecule Uptake by Microorganisms

Soil microorganisms represent a large part of the living biomass but in general are not used for bioaccumulation studies since they cannot be separated from the soil to measure the pollutants. Nevertheless important information could be obtained from an experiment in which antibiotic resistant bacteria with bioaccumulated difocol were introduced into the soil. Radioactive ^{14}C-dicofol was bioaccumulated during the incubation of Pseudomonas fluorescens (soil bacteria) resistant to kanamicyn and rifampycin. These bacteria can be measured in the soil using CFU counts with petri dishes containing the above mentioned antibiotics able to suppress all other soil microorganisms. ^{14}C-dicofol was incubated and bioaccumulated in Pseudomonas fluorescens strains after which it was poured on top of a soil and subjected to a succession of simulated rainfall events. At the end of the simulated rainy period of 24h, 95% of the radioactive labeled insecticide remained in the upper 1cm of the soil, whereas 60% of the microorganisms had been transported 10 cm through the soil and were recovered in the leachate. Only 7% of the bacteria poured onto the soil remained within the first 1 cm of soil. In this experiment the acaricide dicofol moved away from the bacteria toward the soil particles showing that it was more strongly attracted to the soil matrix than to the living cells. Therefore in this case the soil acted as a filter/sponge, protecting microorganisms, mesofauna, plant roots and preventing ground water pollution. Molecules will move to the sites they find most attractive and in this case the soil matrix showed a higher uptake of the pesticide, thus reducing the dicofol content in the bacterial cell envelope. In organic matter rich soils, it is not uncommon for many POP´s to show preferential partition towards soil organic matter adsorption rather than to plants or other living forms. A good parameter to evaluate preferential partition of pollutants between soil and biosphere is to compare octanol/water (K_{ow}) ratios of each compartment. A higher octanol/water ratio (K_{ow}) for organic matter than for the soil bacteria explains why the dicofol moves out of the cell and into the organic component of the soil.

Lipophilic Molecule Uptake by Earthworms

Earthworms, as a "living system" model, facilitate bioaccumulation studies in soil since they can be collected easily and analyzed for pollutant uptake. Papini and Andrea working with simazine, a relatively non-polar (K_{ow} 2-2.3) herbicide, and Paraquat, a highly polar herbicide, found that simazine did not bioaccumulate in the earthworm Eisenia foetida but Paraquat did. This result was the opposite of what was expected from the point of view that non polar substances bioaccumulate in the biosphere and polar substances do not. In a separate but similar study using an Argisol soil, the herbicide atrazine did not bioaccumulate in the earthworms Pontoscolex corethrurus either. However, these results are not fully explained by the polarity of the pesticides. To interpret these results more precisely requires an understanding of K_{ow} as well as knowing the importance of organic matter in the soil. Soil organic carbon content (K_{oc}) correlates positively with soil K_{ow} and is an important factor to attract lipophilic substances. Given that K_{oc} and K_{ow} in general correlate positively, soil organic matter (OM) content can be used to select soils for study. Andréa and Papini used this method to compare how simazine and paraquat would behave in soil in the presence of the earthworm Eisenia foetida with different amounts of soil organic matter.

Two soils, one with 12 g. L^{-1} organic matter (pH 5.7) and the other with 93 g. L^{-1} organic matter (pH 6.4), were treated with simazine and paraquat. Soils were incubated for 90 days with Eisenia foetida and the bioaccumulation coefficient factor was determined. Simazine did not bioaccumulate in the earthworms in the high organic matter soil; however, in the low organic matter soil it did. With simazine, it seems likely that molecular polarity controlled the different distribution of pesticides between soil organic matter and biota. In the soil with low organic matter, preferential partition was toward earthworms that provided more attractive sites for the pesticide than the soil matrix. In the high organic matter soil, preferential partition was toward the soil organic matter, less polar than the earthworms and therefore chemically more attractive to lipophilic pesticides. The results at first appear to be at odds with one another, in terms of where the pesticide accumulated, until we recognize that pesticide movement is not necessarily unidirectional.

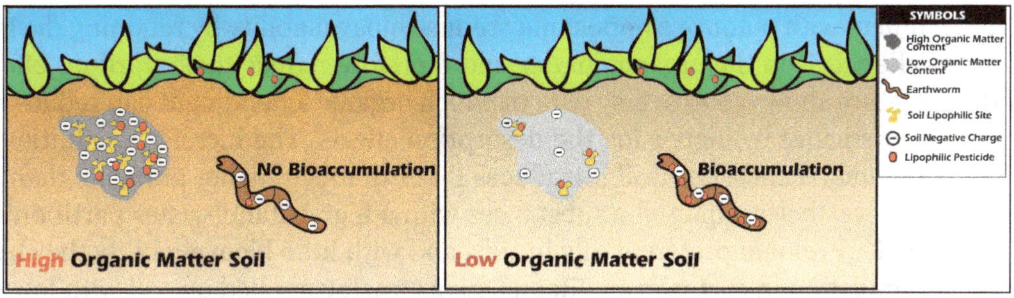

The movement of lipophilic pesticides in soil is influenced by the relative abundance of lipophilic sites in the living and non-living soil components.

Lipophilic Molecule Uptake between Earthworms and Plants

Experiments with p,p`-DDE in soils with the earthworms Ersenia fetida, Lumbricus terrestris or Apporectodea caliginosa in the presence of different plant subspecies showed that Curcurbita pepo ssp. reduces bioaccumulation in the three species of earthworms. Similar experiments in the presence of Curcurbita ovifera ssp. showed a reduction as well as an increase of bioaccumulation in the different earthworm specie. Bioaccumulation in the plant C. pepo were enhanced with the three earthworm species which did not occur with C. Ovifera where only a slight increase was observed. In this set of experiments preferential partition occurs mainly toward plants and less to earthworms.

An increase in the bioaccumulation in plants indicates higher bioavailability of this residue due to presence of the earthworm which through chemolysis is able to change humic acid and increase the protein and carbohydrate moieties and degrade the carboxylic and aliphatic groups. In consequence the hydrophobic index HI = 0.0433 – 0.0811 in the soil decreases in the presence of earthworms and by C. pepo to 0.0231- 0.0286, a condition that reduces soil adsorption and increases bioavailability and therefore bioaccumulation is enhanced in the plant. Other data showed that the phytoextraction capacity of plants is related to the capacity of inorganic uptake from soil. Fertilizer amendment with N and P enhances phytoextraction and increases bioaccumulation in C. Pepo.

Bioaccumulation of lipophilic substances such as chlordane is quite different between plant species, as observed by zucchini with a low and pumpkin with a high uptake. These lipophilic substances are normally bioaccumulated in roots and only a small amount is translocated in a decreasing sequence to stems, leaves and fruits. When K_{ow} is higher than 5, plant uptake is considered to occur mainly via the air-to-plant route. These data agree with the observations of Schnoor et al. that plant uptake is very efficient for moderate hydrophobic organic chemicals with a K_{ow} of 0.5 to 3. For a K_{ow} higher than 3 these chemicals bind more and more strongly to the surface roots with decreasing translocation within the plant. However, translocation of chemicals such as terbuthylazine, with a K_{ow} of 3, and atrazine can occur in high amounts. Fairly soluble chemicals with a K_{ow} lower than 0.5 are not sufficiently sorbed to roots and are not actively transported through plant membranes.

Soil amendment with manure compost may reduce bioavailability by retaining the toxic organic chemicals in the organic matter and therefore reduce the hazardous effects but the literature shows controversy data in which organic amendment can reduce adsorption of pesticides by increasing the desorption effects. The increasing addition of sludge as the final disposal on soil introduces POPs in much higher amounts than air deposition. Nevertheless lipophilic substances with a high octanol-water partition coefficient (log K_{ow}) remain preferentially in soils and with little bioavailability they have low bioaccumulation in earthworms. Radioactive atrazine applied on soil with low organic matter content previously covered by cattle manure, showed a slower leachate speed compared to control but with a low retention capacity in the soil. Soils modulate adsorptions and bioavailability and an inverse correlation occurs with a decrease of bioaccumulation in earthworms when K_{ow} increases, which is different from an aqueous environment when there is a positive correlation between K_{ow} and bioaccumulation.

Cation Bioaccumulation

Many evidences indicate that the lipophilic character of soil organic matter is one of the most important factors for preferential partition of lipophilic substances toward soil; nevertheless this process can be carried out by chemical bonds such as ionic charged bonds of organic toxic chemicals. Below is a description of how preferential partition works between the soil and the living biosphere with cationic charged molecules. Bioaccumulation in terms of cationic charged molecules between the soil and biosphere has not been reported in the literature before.

Based on the hypothesis that polarity is the main factor controlling bioaccumulation, one would expect that a strongly polar pollutant, like paraquat, would be accumulated in the most polar parts of the soil. Thus in soils with low organic matter, one would expect no bioaccumulation in earthworms and paraquat would be bound within the soil matrix. Nevertheless, Papini and Andrea found the opposite. Preferential partition moved this compound towards earthworms, depending on the amount of ^{14}C-paraquat (1.2; 12 and 120 μg. a.i.g^{-1}) applied to the soil. To understand these results we have to note the importance of the relative abundance of charge-binding sites (attractive/exchange sites) in the biosphere and soil. In soils with low OM and low ion exchange capacities, the exchange sites, which sequester positive charged pesticides that gradually become saturated and consequently makes possible simultaneous available pesticides to bind anionic sites on or in earthworms. This is a competition between the earthworms and attractive soil sites for paraquat. In this low SOM soil, paraquat bioaccumulation did not surpass Bcf 5 probably because the anionic sites on the surface of the earthworm cells were limited and already saturated.

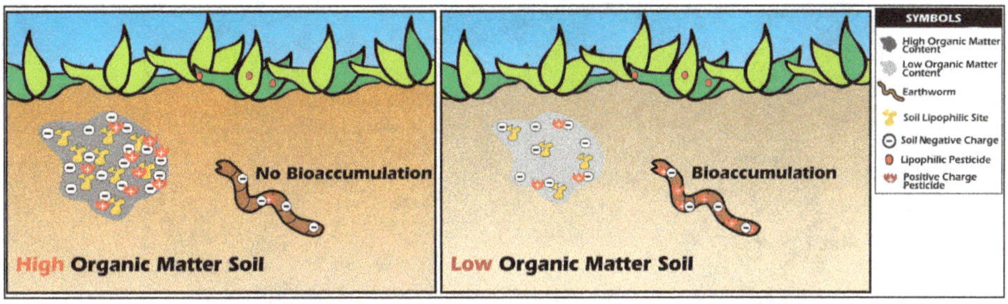

In preferential partition charged pesticides are attracted to anions between the abundance in soil and biosphere.

In a high OM soil and with a low application of paraquat (1.2 μg a.i.g^{-1}), the bioaccumulation factor was 1.1 and increased with higher concentrations of applied paraquat up to Bcf. 3.8. With increasing paraquat concentrations (12 to 120 μg a.i.g^{-1}) one would predict that as the soil charge sites gradually became saturated then gradually more paraquat would become attracted to the earthworms. From these experiments, we noted the predominance of electrostatic binding in the soil and the importance of an abundance of exchange sites. In comparison, lipophilic attraction is driven by affinity without limits of concentrations which are different from the electrostatic bonds were the charged sites involve higher bond energy and therefore are predominant but have quantitative limitations with pesticides up to saturation with consequences in bioaccumulation. The correlation between K_{oc} (soil organic matter) with K_{ow} (molecular polarity) and availability of cation exchange sites (electrostatic bonds) in the soil shows that the Preferential Partition concept can explain pesticide bioaccumulation.

Ecotoxicology depends on soil organic matter. When the SOM is high the ecotoxicological effects are low and when the SOM is low the effects are high. In this latter case the residues remain bio-available and are intensely absorbed in the biosphere and therefore are hazardous to flora and fauna. Polarity and/or anion charge capacity of living organisms compete with the parameters in the soil organic matter. The highest attraction capacity of these different compartments defines the way in which molecules move in the environment. Molecules inserted in the soil matrix by "Preferential Partition" can be protected from hydrolysis, oxidation/reduction reactions as well escape enzymatic action.

Microbial Constraints of Biodegradation

All natural molecules can be biodegraded in the environment. An important constraint of xenobiotic biodegradation is the absence of microorganisms with efficient biodegradation capacity in a specific environment. Recently there has been an intense research effort to develop or transfer microbial efficient biodegradation genes to a microorganism that is adapted to a specific environment but lacks biodegradation capacity. Other constraints can occur when microorganisms do not enter in small soil microspores and therefore could not be present to promote biodegradation.

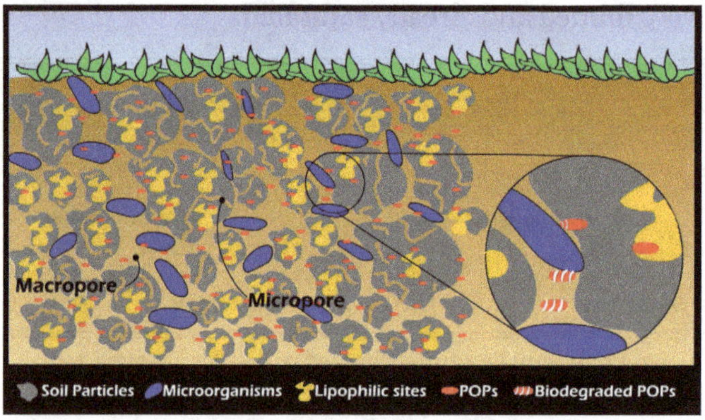

Restriction access of microorganism in the soil by micropores.

Movement of Persistent Molecules in the Environment

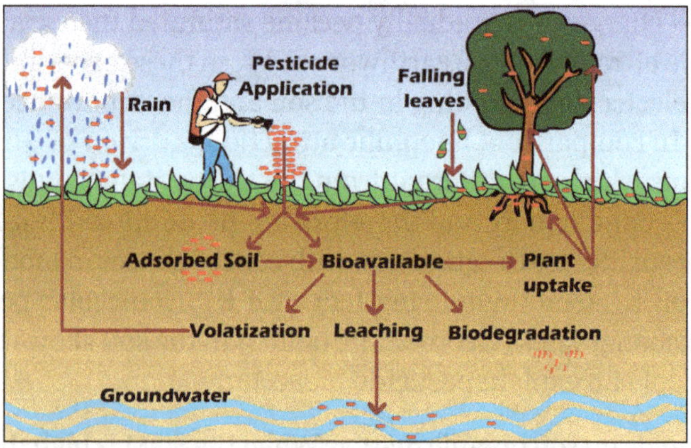

Overview of POP dispersion in the environment of air, water and biosphere.

Desorbed or non sorbed molecules are bio-available and can move into the food chain and can also get into ground/surface water or reach the atmosphere through volatilization and are thus be randomly disseminated in the biosphere. When persistent molecules are adsorbed in the soil they cause less hazardous effects than when they are bioavailable. Biodegradable substances, differently from persistent molecules, cannot survive for a long time in the environment. The problem is not only the actual pollution of pesticide applications, which is strongly dependent on the climatic conditions but also the composts with the organic wastes containing POPs that are added to soils as well the inadequate forms in which POPs used to be stored.

Movement from Soil to Water

The vertical displacement of substances in the soil can leach from the surface through the unsaturated zone into deeper layers down to ground water. It is well-known that the soil infiltration capacity depends on soil texture characteristics, porosity, and humidity. Soils with a sandy texture are more susceptible to the process of leaching while clay soils have greater pesticide adsorption potential. Leaching can occur in an irregular manner known as preferential flow. This is due to irregularities in the soil that facilitate an easy path for water and pollutants at specific points. These paths can be formed by soil cracks, worm trails, rotten roots, termites, and other biological activities. Soil in which surface layer rich in organic matter was removed by flattened to build roads or courtyard enhance strongly the leaching process. Water consumption from wells exposed to pesticide applications can be very dangerous for humans and cattle. Generally, ground water in the dry season drains to rivers and in the raining period receives water from rivers and pollutants follow these processes.

Surface water contamination with persistent molecules occurs not only by the drainage processes from ground water but also strongly increases with runoff. When the quantity of rainfall exceeds the soil infiltration capacity, superficial runoffs occur. Runoffs begin with small lamina flows that build up until they form turbulent flows which cause pesticide losses and even gully erosion. This process has been shown to be an important source of pollution in surface waters. Another source of pollution is the discharge of effluents from waste water treatment plants to surface waters. Pollutant molecules can be distributed by rivers and affect different living forms in the water as well as the fauna that use it for drinking. Persistent molecules contaminating these waters are also spread to agriculture products through irrigation.

Movement in the Air

The sources of POP in the air are pesticide applications, oil combustion, industries and indoors at home. During the application of pesticides a part remains in the air due to drift. Another very important source is volatilization of pesticides from soil or plant surfaces. Pesticide dissipation occurs when the product is being sprayed. Droplets are aerially dispersed and reach non-target areas. The spray drift of pesticides is influenced by the size of the pulverized droplets, application pressure, distance of the pulverizer nozzle to the target and the velocity of the wind during application. By strong volatilization the relative loss of the applied amount can be very high.

These molecules contaminating the air are transported by air currents and redistributed globally, polluting all kinds of ecosystems. Today we may find many such molecules in the Antarctic and in obscure ocean islands. This pollution affects man as well as many animal species. Besides the adsorption capacity that is able to immobilize these molecules, another important variable is the molecular vapor pressure which is influenced by temperature – an increase of 10 °C makes these molecules about four times more volatile. This process is much greater in the tropics with higher temperatures than in temperate regions of the world.

Also these losses can affect people, domestic animals, pollinating insects as well as contaminating lakes, rivers, and fish or other aquatic plants and animals. Air pollution is quite difficult to control. Trees and forests can filter the pollution that is spread by horizontal wind but not by rainfall. Air pollutants can move from rural areas to towns and can be present everywhere. This distribution is determined by the wind and physical barriers such as mountains etc. The challenge is how to reduce this process as much as possible considering that these molecules have been lost for the purpose that they were intended and non target flora and fauna are not able to escape from their hazardous effects efficiently.

Air pollution can also begin indoors and then move outdoors. Polybrominated diphenyl ethers (PBDEs) and phthalates among others are some examples. Brominated compounds make up an important group of flame retardants. Some of these products, such as Penta-BDE, seem to show toxicological effects at very low concentrations. PBDEs are structurally similar to thyroid hormones acting as endocrine disruptors via alterations in thyroid hormone homeostasis and demonstrate neurodevelopmental effects. The most sensitive populations are likely to be pregnant women, developing fetuses, and infants.

In recent years, PBDEs have been recognized as significant pollutants of the indoor environment. These additives are mixed into plastics and foams but do not form chemical bonds. This makes them much more likely to leach out of goods and products by volatilization. PBDEs are lipophilic compounds, and when released into the environment bioaccumulation can occur in living organisms and this is followed by biomagnification in the food chain. Bioaccumulation in wildlife has been reported in numerous studies, even in places with no local sources or industrial production like the Arctic. Due to their high production volume, widespread usage, and environmental persistence, PBDEs have become ubiquitous contaminants in environmental media, biota and humans. As their levels are rapidly increasing in the environment, these chemicals have evolved from 'emerging contaminants' to globally-distributed organic pollutants.

Movement in the Food Chain

Here some aspects of the way in which persistent molecules can be transferred from one life form to other will be emphasized, considering that the overall food chain sequence processes are well known.

Cell Bioaccumulation

How do persistent molecules move into cells in higher concentrations than the surrounding environment? "Preferential partition" describes the "choice" of the substance, due to affinity, between cell membranes or components of the environment. Affinity is characterized mainly by lipophilicity and charge binding. With the photo-binding technique it is possible to identify in which cell component the molecule will preferentially bind. This technique preserves the binding when only weak bonds between the pollutant and cell are available. These weak bonds are not resistant to the harsh conditions of cell fractionation. The method establishes covalent bonds by UV irradiation only between pollutants containing aromatic structures and the cell component where weak bonding occurs and this covalent form of bonding can resist cell fractionation. Studies with the gram negative soil bacteria Azospirillum lipoferum, have shown that dicofol, which is easily hydrolyzed turns persistence by becoming imbedded in the membrane. In this case dicofol was found in the neutral lipids of membranes where it influences the microdomain of membrane bound enzymes, as also occurs with other molecules. This characterizes a non specific mode of action as seen with other molecules that interfere with different membrane bound enzymes such as ATPase and nitrogenase.

Ecotoxicological Effects in Soil Living Organisms

The key message is that ecotoxicological effects are not only defined by the intrinsic chemical nature of each compound but also by the interaction of their effects properties within a given environment. Applying the concept of "Preferential Partition" helps us to understand conflicting reports about bioaccumulation and the ecotoxicological effects in soils. Bioaccumulation is the natural process that gradually concentrates non toxic levels of pollutants into toxic levels within a biota causing unpleasant side effects.

Different side effects have been described for the same pesticide. For example in the nitrogen cycle, some authors reported an inhibition effect of a pesticide while others reported, for the same molecule, an increasing or no effect on the ammonification, nitrification and nitrogen fixation processes. This depends mainly on the differences in the soil used and concentrations applied. "Preferential Partition" explains why substances remain in the soil matrix, when they are likely to be bioaccumulated in living forms or move through the soil to pollute water resources or are suitable for biodegradation.

Non soil bound residues are an essential condition for biodegradation. Persistence occurs mainly when low numbers of or no biodegrading microorganisms are present or most commonly when despite the presence of biodegradation microorganisms, the residues are entrapped and not bioavailable.

Hazardous organic substances that are strongly adsorbed in soil organic matter pose less risk than those with low adsorbance. Sufficient soil organic matter reduces bioavailability and decreases bioaccumulation in the living biosphere but in turn results in persistence with long term pollution in soils. The balance between adsorbed and bioavailable molecules determines the ecotoxicity levels and this is determined by the physic-chemical characteristics of the soil as well as the pollutant molecule. This balance is changed by the weathered pollutants in the soil that have a lower bioavailability than the recently applied products. Therefore it is a pitfall to conclude that the hazardous effects are only caused by pesticide concentrations in the soil.

Plant Uptake and Food Chain up to Fauna

One of the most important sources of pollution in the food chain is soil contaminated with pesticides and other POPs mainly introduced by air deposition. Another way is the direct surface uptake mainly in leaves during application but also by plant uptake powered by plant evapotranspiration in which persistent molecules dissolved in soil water is moved from roots to shoots. From the roots the pesticides move by translocation to stems and then often a strong bioaccumulation occurs in the leaves or fruits. Crops where pesticides are used intensively are consumed by cattle, humans or wild life. A strong increase the concentrations of these molecules can occur in this process called biomagnifications.

Human contamination with pesticide moved by food chain.

In order to minimize ecotoxicity we need to restrict the inappropriate use of pesticides and thereby remove them from the food chain and water reserves. Preferential Partition presents a concept - tool which is able to estimate what happens to pesticides in crop and fodder plants. In soils with high organic matter content, lipophilic and or charged pesticides are retained in the soil organic matter and the uptake into plants decreases. The reverse occurs in low organic matter sandy soils with low cation exchange capacity CEC. Light textured sandy soils do not adsorb and retain in high amount hazardous products and will both bioaccumulate in living tissues and pollute water resources. In soils of this type, plants strongly adsorb pesticides resulting in enhanced contamination with subsequent phytotoxicity and toxicological effects on fauna. This soil type would be a worst case scenario adequate to be the bench mark for pesticide registration. The pesticides that could be used under these conditions can be used generally in all soils. Registration based on strongly adsorbing soils needs to be avoided because it camouflages the eco-toxicological effects that occur when the pesticide is applied on less adsorbent soils.

Food Chain up to Fauna

The food chain sequence is where persistent molecules are first adsorbed by plants which are then consumed by animals that are in turn consumed by predatory species. In the first scenario, the animal serves as a filter with systemic eliminations and the concentration of a xenobiotic that remains present in the consumable product is much lower than the original concentration in a feed material. For meat (including adjunctive fatty tissue), this would mean that in many cases human exposure is very low or even negligible for almost all compounds. The second scenario addresses the accumulation of chemical residues in animal tissues such as liver and kidney with notorious examples including the accumulation of heavy metals, melamine and other nephrotoxins in the kidney, and dioxins and polyhalogenated persistent environmental pollutants (POPs) in the liver and fatty tissues. Persistent molecules increase in each ingestion with the uptake of higher concentrations along the food chain that can result in very poisonous and hazardous concentrations for humans, cattle and wild life. The worst bioaccumulation seems to be salmon produced on fish farms.

The metabolic pathways of digestion in diverse animal species are different. Strict herbivorous species digest soluble fibers in their large caeca (the equivalent of the rumen) with the help of complex microbiological flora. This microbial fermentation may result in the release of bound (plant) toxic products which are often conjugated to sugars (glucosides) in plants and in this form are biologically inactive (or less active). The morphological and physiological characteristics of the gastro-intestinal tract in farm animal species largely determine the rate of absorption of a contaminant. The internal dose of a toxicant that reaches the post-hepatic circulation, the rate of excretion and subsequently the potential carry over and cumulative potential in tissues need to be considered. The most significant diversity in species is observed at the level of hepatic biotransformation expressed by cytochrome P_{450} enzymes and their polymorphisms. Diversity can also be observed in all phase of enzymes such as glutathione-s-transferases, glycine conjugates, UDP-glucuronosyltransferases that are expressed in the intestinal wall and in the liver.

The choice of toxicity endpoints may differ between animal health and human risk assessments. Hence, clinical reports from intoxications in different animals when available may partly provide a basis for hazard characterization whereas other important endpoints maybe applied to human risk assessment as an extrapolation of the dose response from laboratory animal species to humans or using the dose response provided by human epidemiological data. This is typically illustrated with genotoxicity and carcinogenicity data, which are prominent endpoints in human hazard characterization, but which are not commonly used in farm animal health risk assessment considering their relatively short lifespan. The quantitative transfer of a substance from feed to an animal-derived product is commonly expressed as the carry-over rate and serves as a basis to establish maximum tolerable limits (MLs) in animal feed and animal-derived foods.

Chronic Toxicology for Animals and Humans

Bioaccumulation is the natural process that gradually concentrates non toxic levels of pollutants into toxic levels within a biota causing unpleasant side effects. Bioaccumulation involves silent natural processes which stealthily and inevitably affect all biota. Today hazardous substances are ubiquitous, albeit mainly in low concentrations, and for the majority of us they are an unconscious part of our daily diet. The regular intake of sub toxic levels of persistent pollutants can gradually bioaccumulate up to toxic levels and after time produce chronic effects which today are recognized and understood as such.

More recently low concentrations have been shown to cause effects such as endocrine disruptors (EDs) quite different from those observed by high concentrations. EDs are recognized as being partly responsible for a global reduction in male fertility, sexual abnormalities), adrenal function disorders, human obesity and other metabolic disturbances not to mention similar and worse effects on other living organisms. Recently new evidences have shown the effects of POPs (mainly pesticides) on metabolic disturbances related to obesity, insulin resistance promoting diabetes 2 and the association of organochlorine with vitamin D deficiency.

The bioaccumulation of POPs by mothers, over many years, exposures the fetus to these pollutants during pregnancy. Despite the difficulty of experimentation with POPs in humans, follicles could be analyzed when "induced reproduction" occurs and results have shown that contamination in the oocyte follicle can reduce human embryo quality with consequences for future generations caused by the harmful effects of these hazardous chemicals.

Clearly, the threat to the environment and human health is reduced when pesticides are not bioavailable and cannot concentrate in the food chain. Food grown in soils where pesticides are strongly adsorbed has lower health risks.

Cell Mortality and Redistribution of Persistent Molecules

The presence of persistent molecules in different forms of life is temporary limited by plant senescence and/or microbial/fauna death. Two possible routes can occur; one following the food chain and the other by a rotting process in which recalcitrant molecules can be released. In plants the rotting process of wood and straw is a complex process in which the twisted lignin/hemicellulose/cellulose fibers follow a cascade where the first step is the "loosing" of the cell wall. This occurs by non enzymatic peroxidation (Fenton reaction) that changes the lignin structure and reduces the barrier effect of the cell wall allowing the second step of enzyme diffusion of ligninase peroxidase, Mn peroxidase and laccase into areas where polysaccharides can be hydrolyzed. Laccase is a phenol oxidase that has the ability to degrade many persistent xenobiotics. Other lignolytic enzymes could be released and are also able to biodegrade many persistent molecules.

Nevertheless in this rotting process many recalcitrant substances are not biodegraded or only partially and therefore these molecules follow the normal distribution: adsorption in the soil, free moving molecules that can be removed by runoff, leaching, volatilization and uptake in plants, mesofauna or soil microorganisms.

How to Control Hazardous Effects of Persistent Molecules

The rules that enhance molecular persistence and the parameters that make the wide spread distribution of POPs possible are driven by natural factors, out of human control. Nevertheless recently remediation technology has been developed and in many cases has shown good results to remove pollutants. The approach in this text is to point out different possibilities with reference to some excellent reviews. The technology used was based on physical, chemical and biological methods.

The first step of POP remediation is to promote desorption. Physical methods need to remove the pollutants from soils or water in order to destroy the molecules completely by burning. This has a high cost as the incinerator plant needs filter systems to eliminate other pollutants such as dioxins. Physical methods such as electro kinetic (EK) remediation technology use a low-level direct current as the “cleaning agent”, inducing several transport mechanisms (electro-osmosis, electro-migration and electro-phoresis) and electrochemical reactions (electrolysis and electrodeposition). This technology has already proven its value, especially in contaminated fine-grain soils.

Ozon can be used for chemical remediation of pollutants and there are other methods currently being researched. Chemical degradation reactions in the environment depend on water content, pH, temperature and oxygen reduction potential. These conditions are normally not optimum in nature and therefore degradation is frequently low. Physical and chemical remediations are powerful technologies to remediate small areas that have been polluted by industry or transport accidents. For large areas these process are generally considered too expensive.

Biological degradation processes are promoted by living organisms such as animals, plants but mainly microorganisms. The advantages of microbes are the ubiquitous distribution in normal and extreme environments, fast biomass growth, easy manipulation and high diversity of catabolic enzymes. Bioremediation using microorganisms compared to the other methodologies can be used in polluted soils and waters and is less expensive than all other methods. Nevertheless the high costs are limitations for large polluted agricultural areas. One of the most powerful approaches is phytoremediation as some plants have a strong capacity to extract pollutants from the soil and incorporate these molecules into their own biomass. Nevertheless in phytoremediation of hydrocarbon pollutants (HCs) the accumulation between roots and shoots is quite low compared to plant-promoted biodegradation in the rhizosphere. Root exudation stimulates microorganisms in the rhizosphere resulting in enhanced mineralization.

Today efforts are being made to identify the most adequate alternative remediation in a framework of possibilities considering parameters that characterize cost-risk tradeoffs and uncertainty impacts.

BIOMAGNIFICATION

Biomagnification, also known as bioamplification or biological magnification, is any concentration of a toxin, such as pesticides, in the tissues of tolerant organisms at successively higher levels in a food chain. This increase can occur as a result of:

- Persistence: Where the substance cannot be broken down by environmental processes.
- Food chain energetics: Where the substance's concentration increases progressively as it moves up a food chain.
- Low or non-existent rate of internal degradation or excretion of the substance – mainly due to water-insolubility.

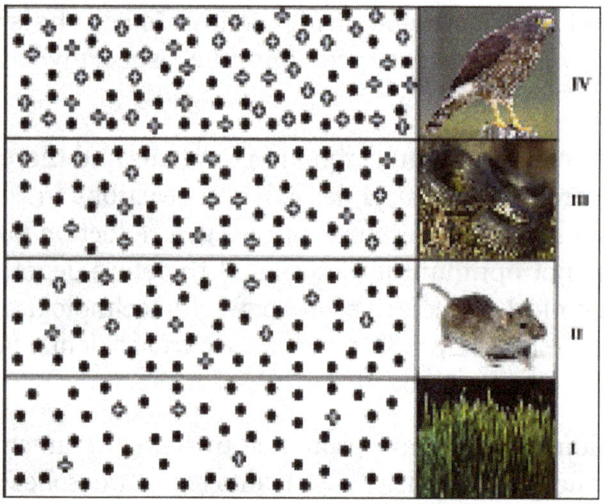

In biomagnification the concentration of the persistent toxins (crosses) increases higher up the food chain.

Biological magnification often refers to the process whereby certain substances such as pesticides or heavy metals work their way into lakes, rivers and the ocean, and then move up the food chain in progressively greater concentrations as they are incorporated into the diet of aquatic organisms such as zooplankton, which in turn are eaten perhaps by fish, which then may be eaten by bigger fish, large birds, animals, or humans. The substances become increasingly concentrated in tissues or internal organs as they move up the chain. Bioaccumulants are substances that increase in concentration in living organisms as they take in contaminated air, water, or food because the substances are very slowly metabolized or excreted.

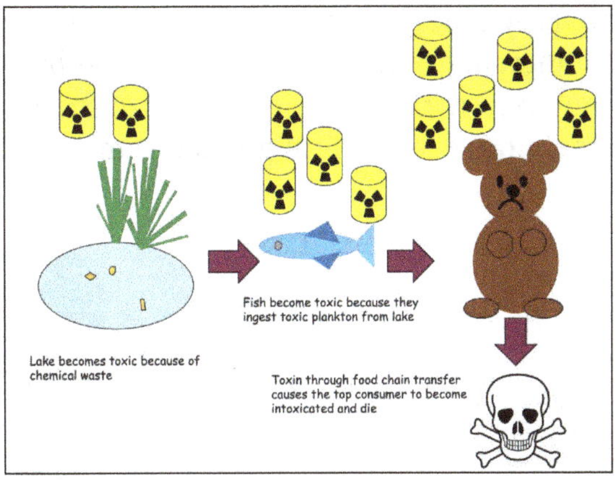

In this scenario, a pond has been intoxicated. As we go further into the food chain, the toxin concentration increases, causing the top consumer to eventually die of intoxication.

Processes

Although sometimes used interchangeably with "bioaccumulation", an important distinction is drawn between the two, and with bioconcentration.

- Bioaccumulation occurs within a trophic level, and is the increase in the concentration of a substance in certain tissues of organisms' bodies due to absorption from food and the environment.
- Bioconcentration is defined as occurring when uptake from the water is greater than excretion.

Thus, bioconcentration and bioaccumulation occur within an organism, and biomagnification occurs across trophic (food chain) levels.

Biodilution is also a process that occurs to all trophic levels in an aquatic environment; it is the opposite of biomagnification, thus when a pollutant gets smaller in concentration as it progresses up a food web.

Lipid, (lipophilic) or fat soluble substances cannot be diluted, broken down, or excreted in urine, a water-based medium, and so accumulate in fatty tissues of an organism, if the organism lacks enzymes to degrade them. When eaten by another organism, fats are absorbed in the gut, carrying the substance, which then accumulates in the fats of the predator. Since at each level of the food chain there is a lot of energy loss, a predator must consume many prey, including all of their lipophilic substances. For example, though mercury is only present in small amounts in seawater, it is absorbed by algae (generally as methylmercury). Methyl-mercury is the most harmful variation of mercury. It is efficiently absorbed, but only very slowly excreted by organisms. Bioaccumulation and bioconcentration result in buildup in the adipose tissue of successive trophic levels: zooplankton, small nekton, larger fish, etc. Anything which eats these fish also consumes the higher level of mercury the fish have accumulated. This process explains why predatory fish such as swordfish and sharks or birds like osprey and eagles have higher concentrations of mercury in their tissue than could be accounted for by direct exposure alone. For example, herring contains mercury at approximately 0.01 parts per million (ppm) and shark contains mercury at greater than 1 ppm.

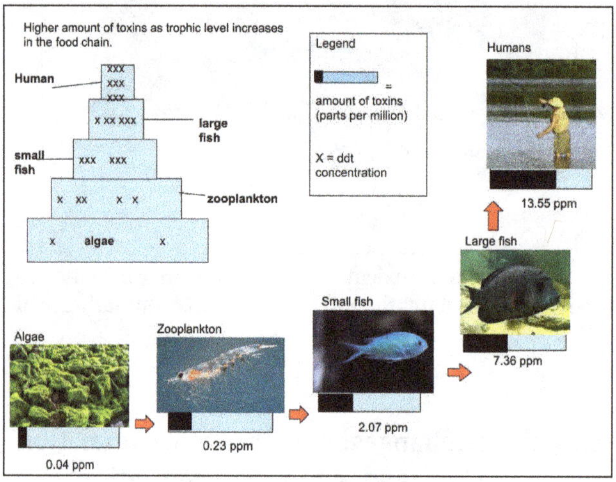

Biomagnification is the build up of toxins in a food chain. The DDT concentration is in parts per million. As the trophic level increases in a food chain, the amount of toxic build up increases. The x's represent the amount of toxic build up accumulating as the trophic level increases. Toxins build up in organism's fat and tissue. Predators accumulate higher toxins than prey.

DDT is thought to biomagnify and biomagnification is one of the most significant reasons it was deemed harmful to the environment by the EPA and other organizations. DDT is stored in the fat of animals and takes many years to break down, and as the fat is consumed by predators, the amounts of DDT biomagnify. DDT is now a banned substance in many parts of the world.

A large number of studies, Suedel et al. concluded that although biomagnification is probably more limited in occurrence than previously thought, there is good evidence that DDT, DDE, PCBs, toxaphene, and the organic forms of mercury and arsenic do biomagnify in nature. For other contaminants, bioconcentration and bioaccumulation account for their high concentrations in organism tissues. More recently, Gray reached a similar substances remaining in the organisms and not being diluted to non-threatening concentrations. The success of top predatory-bird recovery (bald eagles, peregrine falcons) in North America following the ban on DDT use in agriculture is testament to the importance of biomagnification.

Substances that Biomagnify

There are two main groups of substances that biomagnify. Both are lipophilic and not easily degraded. Novel organic substances are not easily degraded because organisms lack previous exposure and have thus not evolved specific detoxification and excretion mechanisms, as there has been no selection pressure from them. These substances are consequently known as "persistent organic pollutants" or POPs.

Metals are not degradable because they are elements. Organisms, particularly those subject to naturally high levels of exposure to metals, have mechanisms to sequester and excrete metals. Problems arise when organisms are exposed to higher concentrations than usual, which they cannot excrete rapidly enough to prevent damage. Some persistent heavy metals are especially dangerous and harmful to the organism's reproductive system.

Food Web-Specific Biomagnification

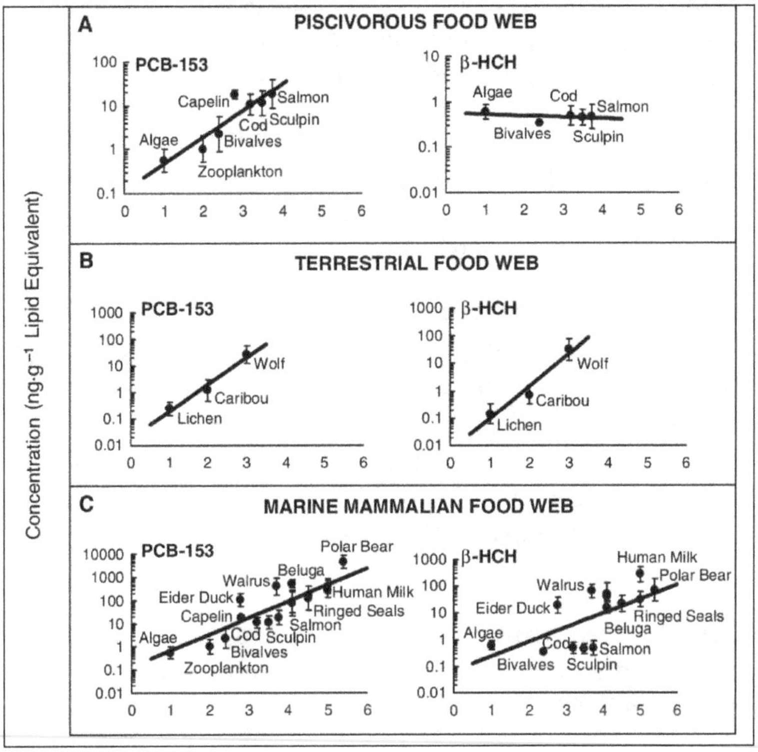

To identify bioaccumulative substances, regulatory authorities rely on the chemical's K_{ow} or, when available, on organism/water chemical concentration ratios measured in laboratory tests [bioconcentration factors (BCFs)] or in the field (bioaccumulation factors). These criteria were derived from laboratory tests in fish, which revealed correlations between the K_{ow} and the BCF of organic chemicals and showed that bioconcentration could be explained and predicted by lipid-water partitioning. Studies in real food webs demonstrated that bioaccumulation in food webs is not solely a lipid-water partitioning process. Dietary accumulation or biomagnification can cause additional bioaccumulation, resulting in an increase in chemical concentration with increasing trophic level in food webs. This apparent chemical transport against the thermodynamic gradient is due to food digestion and absorption, which concentrate ingested chemicals in the gastrointestinal tract. Poorly metabolizable, hydrophobic substances with $K_{OW} \geq 10^5$ have proven particularly susceptible to biomagnification in fish, whereas chemicals with lower K_{OW} do not generally biomagnify in fish.

However, less hydrophobic chemicals (e.g., chlorobenzenes and lindane), with $K_{OW} < 10^5$ and with BCFs in fish-based experiments below the regulatory criterion of 5000, were found to biomagnify in lichen-caribou-wolf food chains in northern Canada. Also, perfluorinated sulfonic acids such as perfluorooctane sulfonate, which (with a calculated $K_{OW} < 10^5$) do not biomagnify in laboratory tests with fish, show a high degree of biomagnification in birds and mammals. These findings indicate that very hydrophobic chemicals with a $K_{OW} \geq 10^5$ are not the only chemicals with biomagnification potential and that lipid-water partitioning cannot serve as a universal model for identifying bioaccumulative substances in wildlife and humans.

Relationship between observed tissue residue concentrations ($ng \cdot g^{-1}$ lipid equivalent) and trophic level for PCB 153 (a high K_{OW}–high K_{OA} compound) and b-HCH (a low K_{OW}–high K_{OA} compound) in Arctic organisms of the piscivorous (A), terrestrial (B), and marine mammalian (C) food webs. Data represent geometric means ± 1 SD.

To test the applicability of current regulatory criteria for identifying bioaccumulative substances, we measured and compiled concentrations of organic contaminants of varying hydrophobicity and K_{OW} in a piscivorous food web (water-respiring organisms only), a terrestrial food web (air-breathing organisms only), and a combined marine mammalian food web (including water-respiring and air-breathing organisms) from northern Canada. We also used mechanistic food web bioaccumulation models to determine the influence of K_{OW} on chemical bioaccumulation in these food chains.

In field studies, we observed strong positive relationships ($r^2 > 0.8$, $P < 0.001$) between trophic level and concentrations of recalcitrant PCB congeners 138, 180, and 153; mirex; hexachlorobenzene (HCBz); dieldrin; and DDTs in all three food webs.

These and related findings support the current regulatory and scientific paradigm that very hydrophobic organic chemicals with a $K_{OW} \geq 10^5$ can biomagnify in food webs. However, although less hydrophobic compounds ($K_{OW} < 10^5$) such as α, β, and γ hexachlorocyclohexanes (HCHs) ($K_{OW} = 10^{3.8}$), tetrachlorobenzenes (TeCBz) ($K_{OW} = 10^{4.1}$), and endosulfans ($K_{OW} = 10^{3.7}$) did not biomagnify in the piscivorous food web, they showed a high degree of biomagnification in the lichencaribou-wolf food chain and in air-breathing organisms of the marine mammalian food web. The predator-prey biomagnification factors or BMFs of β-HCH in ringed seals of 20 and in beluga whales of 50 exceed the BMFs of PCB 180 in those animals. These findings demonstrate that although substances with a K_{OW} below 10^5 cannot biomagnify in fish, they can biomagnify in birds and mammals.

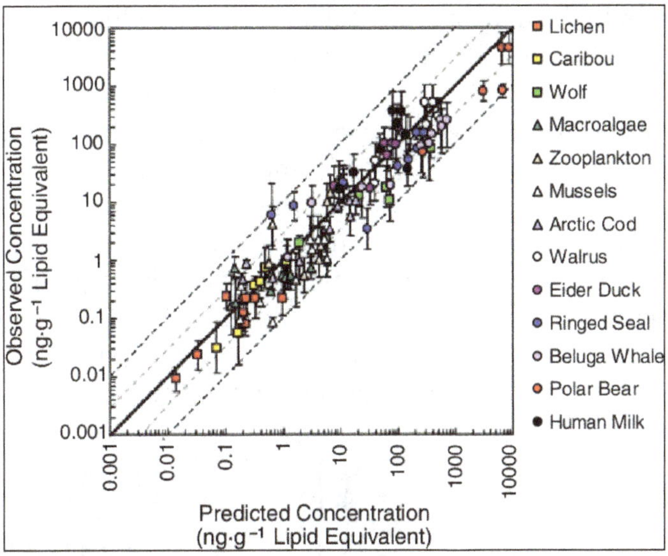

Model predicted versus observed concentrations (ng·g^{-1} lipid equivalent) in various organisms of the Canadian Arctic terrestrial and marine ecosystems. Observed data represent geometric mean ± 1 SD. The solid black line represents an ideal model fit (1:1 predicted:observed line). Dashed lines represent a factor of 3 (gray dashes) and 10 (black dashes) above and below the ideal fit.

Bioaccumulation modeling studies showed good agreement between calculated and observed concentrations of recalcitrant substances in all three food webs. The model shows that air-breathing organisms in this analysis exhibit higher BMFs than those in water-respiring organisms because of their greater ability to absorb and digest their diet, which is related to differences in digestive tract physiology and body temperature. The model also shows that the relationship between the BMF and chemical properties is controlled by the rate of elimination. In water respiring organisms, elimination becomes sufficiently slow to cause biomagnification if the K_{OW} of the chemical exceeds ~10^5. In the airbreathing organisms of this study, this occurs for chemicals with a high K_{OA} ($\geq 10^6$), which causes slow respiratory elimination, and a $K_{OW} > 10^2$, causing slow elimination in urine or nitrogenous wastes. The difference in biomagnification behavior between air-breathing and water-respiring organisms implies that, for substances with a $K_{OA} \geq 10^6$ and a $K_{OW} > 10^2$, K_{OW} and the BCF in fish are not good predictors of biomagnification in air-breathing animals.

Application of the bioaccumulation model to identify potentially bioaccumulative substances among commercial chemicals reveals distinct differences in the biomagnification behavior of chemicals in different food webs. In the piscivorous food web, concentrations of nonmetabolizing chemicals with K_{OW} between 10^5 and 10^8 biomagnify in top-level predatory fish up to 100-fold. No biomagnification occurs for less hydrophobic chemicals with $K_{OW} < 10^5$, which are efficiently eliminated by respiration, or for superhydrophobic organic substances with $K_{OW} > 10^8$, which are absorbed at very slow rates. However, in the marine mammalian food web, which includes water-respiring invertebrates and fish and air-breathing birds and mammals, poorly metabolizing chemicals with a $K_{OW} \geq 10^5$ and $K_{OA} \geq 10^6$ biomagnify, attaining concentrations in top predators (polar bears) up to 10,000 times the concentrations in primary producers. Less hydrophobic chemicals with $K_{OW} < 10^5$ and $K_{OA} \geq 10^6$ also biomagnify strongly, with concentrations in polar bears exceeding those in primary producers up to 3000-fold. Chemicals with $K_{OW} < 10^2$ do not biomagnify in this food web regardless of their high K_{OA} because airbreathing animals eliminate them through urinary excretion. In terrestrial food webs, chemicals with a K_{OW} between 10^2 and 10^{10} and a $K_{OA} \geq 10^6$ can biomagnify up to 400-fold if not metabolized. Chemicals with a K_{OW} between ~10^3 and 10^9 achieve a similar degree of biomagnification, given the same K_{OA}.

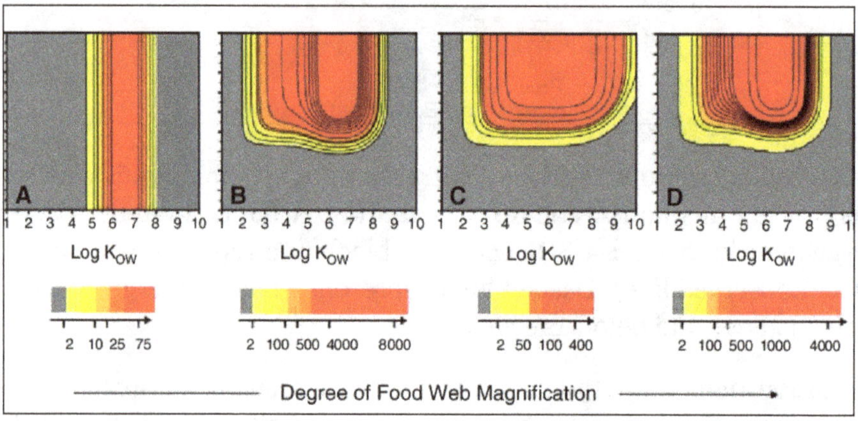

Contour plots illustrating the relationship between chemical K_{OW} (x axis), K_{OA} (y axis), and food web magnification (z dimension represented as contours) in the aquatic piscivorous food web (A), marine mammalian food web (B), terrestrial mammalian food chain (C), and Arctic indigenous human food chain (D). The data represent the combined magnification of the chemical concentrations (ng·g^{-1} lipid equivalent) in the top predator over the concentrations at the base of the food web [e.g., primary producers at trophic level (TL) = 1 to polar bear at TL = 5.4]. A matrix table was generated with ~30,000 K_{OW}-K_{OA} combinations over a log K_{OW} range of 1 to 10 and a log K_{OA} range of 3 to 12. These data demonstrate the combined effect of K_{OW} and K_{OA} on chemical bioaccumulation.

Simulations, representing human dietary exposure of contaminants to the indigenous Inuit population of northern Canada, show that chemical amplifications can reach ~4000-fold for chemicals with a log $K_{OW} \geq 5$ and a log $K_{OA} \geq 6$ and 2000-fold for chemicals with a log K_{OW} between 2 and 5 and a log K_{OA} above 6. The modeling results are consistent with empirical observations for PCBs, chlordanes, HCBz, dieldrin, and DDTs, which have both a high log $K_{OW} \geq 5$ and a high log $K_{OA} \geq 6$ and biomagnify in all three Arctic food webs studied. The results are also consistent with observations of HCHs, TeCBz, and endosulfans, which have a low log $K_{OW} < 5$ but a high log $K_{OA} \geq 6$ and exhibit no biomagnification in the piscivorous food web but biomagnify in air-breathing organisms.

Organic chemicals with a $K_{OW} > 10^2$ and a $K_{OA} \geq 10^6$ (i) have an inherent biomagnification potential in air-breathing organisms of terrestrial, marine mammalian, and human food chains and (ii) include almost two-thirds of all organic chemicals used in commerce. About 40% of chemicals with these properties have a $K_{OW} > 10^5$ and are currently recognized as potentially bioaccumulative because of their high degree of lipid-water partitioning. The remaining 60% of these chemicals, which include substances with a K_{OW} between 10^2 and 10^5 and a $K_{OA} \geq 10^6$, should also be considered for their bioaccumulative potential because of their high degree of lipid-air partitioning. These low K_{OW}– high K_{OA} chemicals are susceptible to biomagnification in air-breathing animals, including humans, because of their slow rate of respiratory elimination. Metabolic transformation can reduce or eliminate the anticipated biomagnification potential but only if the metabolic transformation rate is sufficiently high. In those cases, the bioaccumulation behavior of resulting metabolites should also be considered.

References

- Bioaccumulation, agricultural-and-biological-sciences: sciencedirect.com, Retrieved 16 April, 2020
- Persistence-and-bioaccumulation-of-persistent-organic-pollutants-pops, applied-bioremediation-active-and-passive-approaches: intechopen.com, Retrieved 10 August, 2020
- “DDT Ban Takes Effect”. United States Environmental Protection Agency. 1972-12-31. Archived from the original on 2014-08-12. Retrieved 2014-08-10
- Food-Web-Specific-Biomagnification-of-Persistent-Organic-Pollutants-6211084: researchgate.net, Retrieved 19 June, 2020

CHAPTER 4

Analysis and Measurement

The analysis and measurement of organic pollutants include the analysis of remote zones air quality, total organic carbon, chemical oxygen demand, biochemical demand, etc. All the methods related to the analysis and measurement of organic pollutants have been carefully analyzed in this chapter.

The process to determine the concentrations of the POPs in a sample include a series of events as outlined in figure.

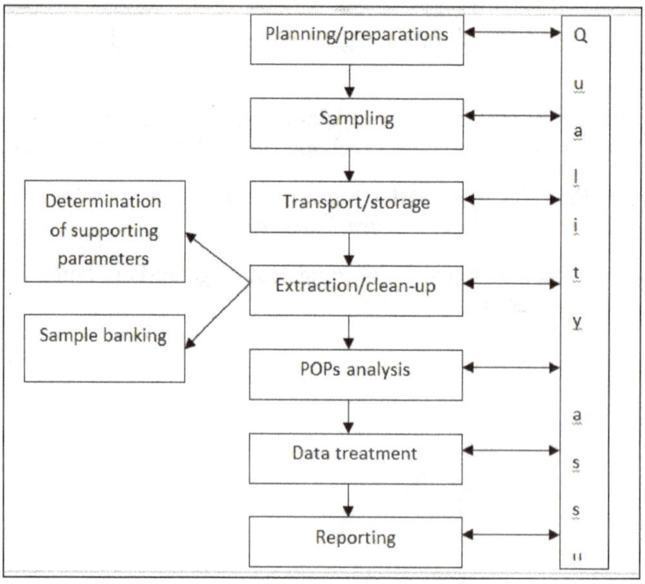

The different stages of the process to determine POPs concentrations.

Sampling

Quality requirements in terms of equipment, transportation, standardization, and traceability are indispensable. It is important that all sampling procedures are agreed upon and documented before starting a sampling campaign.

The analyte, matrix, sampling site, time or frequency, and conditions should be determined depending on the objective of the sampling. In case of human samples it may also be necessary to use a suitable interview form. Although it may be too expensive to get full accreditation for sampling, Quality Assurance and Quality Control (QA/QC) procedures for sampling should be put in place.

No general recommendation can be given with respect to who should perform the sampling. For certain matrices, e.g., human blood, there is no doubt that a specialist, i.e., medical doctor or nurse, has to take the sample. In addition, for human samples, ethical considerations have to be respected. There are pros and cons for sub-contracting a laboratory specialist in sample taking. Sub-contracting the sampling can be an advantage to the laboratories that don't have the required personnel and equipment, but the laboratory must be sure that the sampling was taken under established quality assurance and quality control (QA/QC) conditions. In case an external organization will be sub-contracted to take the sample, it is recommended that the analytical laboratory establishes and provides the sampling protocol. Those in charge of the sampling process must apply security seals, as well as follow the preservation criteria to guarantee the integrity of the sample during transportation.

Extraction and Clean-up

The appropriately prepared sample can be extracted by any one of a number of techniques. The main points to consider are to allow adequate time of exposure of the solvent system in the sample matrix and to limit sample handing steps, i.e. avoid filtration steps by using Soxhlet or semi-automated systems. Extractions can also be accelerated by the use of ultrasonication. Cross contamination from residues left behind by high levels of POPs in other samples is a concern at this stage and equipment must be thoroughly cleaned and checked from batch to batch.

Purity of extraction solvents is also a major consideration. Only high purity glass distilled solvents should be used. Internal standards should be added to the sample as early as possible in the process.

If the results are reported on a lipid weight basis, the determination of the lipid content in the sample is critical. From this aspect the choice of solvents is crucial. If the whole sample is not used for the extraction, the remaining part can be frozen and stored for future control analysis, or analysis of other substances. Likewise the extracts not used in the analysis can be stored, preferably in glass ampoules, at -20 °C.

Isolation steps can be relatively straightforward for low lipid samples such as air. Generally small Silica gel or Florisil columns (either prepared in the lab or pre-purchased) will suffice. The purpose of this step is to remove co-extractive pigments and to separate non-polar PCB (plus p,p'-DDE) from more polar POPs (HCH, most chlordanes, dieldrin/endrin). This is achieved by applying the extract in a small volume of non-polar solvent and fractionating by eluting with hexane followed by one or two other elutions of increasing polarity. Alumina is not recommended because of possible dehydrochlorination of some POPs, e.g. 4, 4'-DDT.

For the human samples a lipid removal step must be included. This can be achieved using size exclusion or gel permeation chromatography (GPC) either in automated systems, using high pressure liquid chromatography (HPLC) columns or by gravity flow columns. The advantage of GPC is that it is non-destructive while the disadvantage is a requirement for large volumes of solvent (low pressure or gravity systems) or expensive columns (HPLC). Lipid removal using sulfuric acid washing or sulfuric acid – silica columns is also effective but does result in loss of some analytes such as dieldrin.

Following fractionation on silica or Florisil final extracts are prepared in small gas chromatography (GC) vials for analysis. Addition of a recovery standard to check solvent volume is recommended at this stage. Careful evaporation is required at this step and only high purity compressed gas (usually nitrogen) should be used.

Analytical methodology for PCDD/PCDF and PCB with TEFs differs from those used for routine ortho-PCB and OCPs in that it requires much lower detection limits (typically 10-100 times lower) because guideline limits in food products are in the low pg/kg range, the Provisional Tolerable Monthly Intake being 70 pg/kg body weight (Joint FAO/WHO Expert Committee on Food Additives. To enforce and control these low concentrations for PCDD/PCDF isotope dilution MS (^{13}C-surrogates for all PCDD/PCDF homologue groups), enrichment on carbon to isolate planar compounds, very small final volumes (10-50 μL) for GC-HRMS quantification is used. Methodology for PCDD/PCDF, slightly modified to include the dioxin-like PCB, developed by the US EPA, is well established and validated by numerous inter-laboratory comparisons. This methodology would be recommended for use in a global monitoring programme. Unlike the guidelines for PCB and OCPs, this very specific guidance for the extraction, isolation and quantification steps for PCDD/PCDF is recommended in order to be in compliance with ongoing programmes and compatible with results generated with these methods over the past 10 years.

Pops Analysis

Since the 1960s, POPs have been determined using gas chromatography (GC) techniques with electron capture detection (ECD), initially using packed columns. Today the separation has been improved by the use of capillary columns and the selectivity by the use of mass spectrometric detectors (MS). Based on the availability of commonly used instruments for the determination of POPs, three types of laboratories can be identified.

Table: Requirements for the instrumental analysis of POPs.

Laboratory tier	Equipment	Infrastructure needs	Chemicals
3	Basic sample extraction and clean-up equipment, capillary GC/ECD	Nitrogen/air conditioning/ power/ personnel specifically trained to operate and troubleshoot equipment problems	Most PCB and all OCPs except toxaphene

2	Sample extraction and clean-up equipment, capillary GC/LRMS	Helium/air conditioning/ consistent power/ personnel specifically trained to operate and trouble-shoot equipment problems	Most PCB and all OCPs; toxaphene if negative chemical ionization is available
1	Sample extraction and clean-up equipment, capillary GC/HRMS	Helium/air conditioning/ consistent power/high operational costs /personnel specifically trained to operate and troubleshoot complicated instrumentation	PCDD/PCDF, all PCB, all OCPs except toxaphene

It is anticipated that improved analytical methods will be developed over the life of the GMP, and the project should be structured so that these improved techniques can be adopted. There is a need to improve the accuracy and lower the costs of these analyses. Emerging procedures with low environmental impact may become more widely available and accepted. It will be necessary to consider comparability as new methods are developed. This could be achieved by analysis of archived samples and direct comparison of new and old methods. Many laboratories are not currently permitted to analyze human blood and milk samples. Special training will be necessary to handle these samples, considering the danger of infectious diseases.

Quality control and quality assurance are important factors in sampling and analysis. Any method performance must be verified through control tables where optimal operational ranges are defined, and the periodical analysis of certified reference materials, own laboratory reference materials, and blind or divided samples should be included in routine QA/QC. The inter-calibration exercises are an essential component in quality assurance of the results and are deemed indispensable in the implementation of a regional laboratory network. A recommendation would be that at least once a year such an intercalibration study is performed for each matrix and persistent organic pollutant of interest to the Region.

Numerous analytical approaches are available for quantifying PCB, and OCPs, as well as PCDD/PCDF by gas chromatography. As with extraction/separation steps only general guidance is required for ortho-substituted PCB and OCPs. Some general guidance on the application of gas chromatographic analysis of ortho-substituted PCB and OCPs is provided in Table for PCDD/PCDF and PCB with TEFs, quantification solely by isotope dilution HRMS is recommended and details can be found in standard operation procedures (SOPs).

HRMS can also be used, of course, for determination of all PCB, including congener-specific determination of non-ortho and mono-ortho substituted PCB as well as

OCPs and indeed would provide a very high level of confidence in the results compared to GC-ECD. However, use of GC-ECD is recommended because of wide availability, relatively low cost, and the substantial knowledge base that exists on the use of this technology for analysis of non-ortho and mono-ortho PCB and OCPs at low ng/g levels or higher in environmental matrices.

Data Treatment

There are a number of parameters that have to be reported together with the analytical results. These include the efficiency of the extraction and clean-up, and the blank values, but the results should not be compensated for these parameters. The uncertainty of the results should also be at least estimated, but preferably determined, using results from inter- or intralaboratory comparisons.

The lowest concentration at which a compound can be detected (limit of detection, LOD) is defined as that corresponding to a signal three times the noise. The lowest concentration that can quantitatively be determined (limit of quantification LOQ) is three times higher than LOD. Compounds found at levels between LOD and LOQ can be reported as present, or possibly as being present at an estimated concentration, but in the latter case the result has to be clearly marked as being below LOQ. Results below the detection limit are often reported as <"LOD".

There are, however, several statistical techniques for treating censored data when the true detection limit is known, e.g. by using a robust statistic such as the median which is unaffected by small numbers reported as below LOD.

Another method uses an estimate of each unknown concentration based on the empirical expected order statistic. This method fits a log-linear regression of the ranked detected concentrations on rank, and then uses this relationship to predict the value of those concentrations reported as below the limit of detection.

Results may also be reported as being in the interval between a value where the lower limit is based on non-quantifiable peaks set to zero and an upper limit where results below LOQ are set as equal to the LOQ. In the analysis of complex mixtures, such as PCB, there is always a risk for coeluting peaks in the gas chromatograms, and known interferences should be reported.

Organization of Quality Control

The organization of the quality control needs special attention in the GMP. Many of the recommendations mentioned above deal with QA/QC measures, but to be able to compare results from different labs and regions (and also for the same laboratory over time) there is a need for overarching activities, such as:

- Distribution of standard materials,

- Organisation of intercalibrations,
- Production of reference materials.

METHODS OF ANALYZING

The 10 POPs substances are classified into three groups for the convenience of analytical operations: i) PCBs, ii) HCB, DDTs, chlordane, and dieldrin, and iii) aldrin, endrin, heptachlor (including heptachlor epoxide) (although mirex and toxaphene are included in the group iii), these substances are excluded from regular monitoring for the time being). Samples taken will be divided into three parts and analyzed. Such classification and way of dividing samples are subject to change according to the discussion and development of analytical methods.

Method of Sampling

Collection of Air Samples

The basic method should be to collect air samples with a combination of high-volume sampler (HV) and quartz filter plus polyurethane foam (PUF, 2 sets) for duration of 24 hours in the volume of 1,000 m^3, as is the case with the collection of PCDDs. However, highly volatile HCB, heptachlor, aldrin and low chlorinated PCBs are subject to breakthrough in this condition, so a combination of low-volume samplers (LV) and two sets of styrene-divinylbenzene type adsorbents (e.g., PS-Air) should be adopted to collect more than several cubic meters of samples concurrently for a duration of 24 hours for analysis. The collection should be repeated three times and the collected matter mixed into one sample, whose extracted solutions should be divided into three equal parts. If the filter has not become clogged after three consecutive days' collection because the air contains little particulate matter, the collected matter may be mixed into one sample, whose extracted solution should be divided into three equal parts for analysis. More detailed collection conditions and pretreatment of samplers will be mentioned separately.

As our primary purpose is to devise a national means, it is important to avoid sampling points adjacent to specified sources such as chemical plants, chop shops and incinerators (setting a sampling point in a large-scale industrial district as a category is another issue), and to select points that allow long-term monitoring. Candidate sampling points should be listed and classified into the following four categories: i) industrial districts, ii) urban areas, iii) suburbs (rural areas) and iv) isolated islands or mountainous areas, and then one point per prefecture should be selected among them in consideration of relations with other candidates.

In collecting air samples, ^{13}C labeled standard sample of the substance to be collected and analyzed should be added to samplers or quartz filters as a surrogate in a known quantity before collection is started. After collection is finished, the entire sample collected is subjected to pretreatment and analysis, and the results are corrected by the recovery rate found through the collection and pretreatment processes. Such recovery rate should desirably fall between 50-120%. For the above-mentioned substances that are subject to breakthrough, the data beyond the lower limit are picked up with first priority from among those whose recovery rates fall within the range of those collected with HV and LV samplers. If there is no appropriate equipment for LV samplers and a sampler is made by yourselves using a pump with equivalent suction power, it should be devised so that the actual total air passage during collection can be measured or estimated accurately.

Sampling should be conducted twice, once in summer and once in winter (by the end of December for winter) by a round-the-clock system. The timing, duration and frequency of each collection are subject to change as a result of the precise surveys of variations during a year to be implemented separately.

Water Samples

Collection should be performed by setting sampling points downstream of major rivers or lakes/marshes, taking water samples in units of 10L, and passing the samples through a C_{18} type sampling filter (e.g., Empore filter). In passing water for sample collection, ^{13}C labeled compound standard solution is added as a surrogate in the same manner as air sampling. This operation is repeated three times or collection is performed at three points, and the samples collected are subjected to extraction. Such extracted solutions are mixed and divided into three equal parts, and then subjected to analysis for each of the three groups mentioned above. If on-site operation is difficult, it is allowable to take accurately 10 liters of water sample in a container with a capacity of more than 10 liters and carry it to an appropriate place, and then pass it through a filter to collect samples. In that case, it is necessary to add a surrogate to the water in the container just after sampling, and perform sampling operation as early as possible, preferably within the day. If the filter becomes clogged during water passage due to high contents of suspended matters, the collecting operation should be performed as follows: Shake well after adding a surrogate, pass it through a glass fiber filter while taking care that suspended substances do not remain on the wall or bottom of the container (washing with purified water when necessary), and further pass it through a C_{18} filter to collect the dissolved substances.

For coastal seawater, around five sampling points should be set in combination with those for bottom sediments every year. A large-volume collection method to be specified separately should be employed, and water samples taken in units of several tens of liters to a thousand liters (1,000L). ^{13}C labeled compound is also added in a known quantity in the same manner as above and the recovery rate corrected. Sampling points are changed every year, and returned to their original points every five years, resulting in around 25 points across the whole country.

Bottom Sediment Samples

In view of the fact that relatively volatile substances are contained in such samples, bottom sediment samples should be subjected to analysis in the wet state without drying. Around five sampling points should be set in coastal areas every year and sampling should be performed concurrently with the coastal seawaters. With this system, it takes five years to cover Japan. Sampling methods should correspond to those employed for the environmental survey on chemical substances. Surface layer sediment is taken at three points near the respective sampling places, and after foreign matter has been removed by visual inspection, the sediment is filtered through a sieve with holes of 1 mm in diameter (16 mesh). Samples are then taken in equivalent quantities and mixed well, and the resulting sample solution is used for extraction. Bottom sediment should not be sampled at places subject to local environmental effects such as inner sections of small bays, places adjacent to dredging and landfill areas.

The sampling locations should desirably be those where the sediment environment is stable and sampling is possible every few years on a long-term basis. Places where the surface layer of sediment may be removed due to the influence of swollen rivers following heavy rainfall, where sediment may suddenly increase, or where the state of sediment in the seashore shows significant regular seasonal changes due to tidal influence, should be avoided. Places where sediments are expected to have been formed over the past year or few years should be selected, together with corresponding collection methods.

While samples are pretreated in the wet state as mentioned above, part of the sample after homogenization should be separately dried using an ordinary method to find its moisture content.

Wildlife Samples

Targeting bivalves, fish and birds that have been sampled for the investigation of Chemicals in the Environment, the number of sampling places should be increased. In particular, sampling places for the bivalves in the coastal survey will be increased while newly targeting green mussels and purplish bifurcate mussels in addition to the present subjects, common blue mussels and hard-shelled mussels belonging to mussels. In those southern areas where these bivalves such as the family of mussels are not found, Japanese oyster or Saxostrea mordax of the oyster family should be added for sampling.

The sampling methods and number of samples should follow those used for wildlife monitoring. However, when frozen bivalves in the shell are defrosted, their blood and bodily fluid are prone to bleed due to the breakdown of tissues. Bivalves taken should be shucked while alive, so that they can stay alive for a few days if kept at low temperatures without freezing in the shell. Nevertheless, shucking should be performed in a day or two since metabolism or excretion is likely to take place at a gradual pace, in which case, a care should be taken not to contaminate with the source in the circumference of working place. It should be noted that the concentration of POPs is reported on a freeze dry weight basis or on a fat weight basis as well as on a tissue wet weight basis in some cases. To allow for the possibility of inter-comparisons, moisture content and fat content should be measured during analytical work using part of the homogenized sample.

For inland water fish, the number of sampling places should be increased. In addition to Japanese dace that is subjected to the present monitoring, bass (Black bass) and bluegill should be included. If any wildlife other than fish can be caught at the specific places every year, coastal squids (one of the family of spear squids) will be a candidate. In the case of squids, their liver organs are to be monitored, and it is acceptable to keep squids frozen in whole and to defrost them and remove the organ just prior to analysis because squids lose very little bodily fluids during defrosting. Attention should be given to the fact that the liver of a squid has a high fat content, and if the homogenized mixture of a sample is left as it is, or is subjected to repeat freezing and defrosting, the fat and the tissue will probably separate and lose homogeneity.

Methods of Measurement

Of the 10 POPs substances, PCBs, DDTs, chlordanes and toxaphene actually exist as a mixture containing multiple constituents. To determine the concentration of each constituent, it is necessary to identify the constituent to be analyzed. For PCBs, the measurement methods should follow the conventional ones in principle. A ^{13}C labeled surrogate standard substance is added in number corresponding to the chlorine number in each of monochloro- to decachlorobiphenyl, and the concentrations of all of the 209 isomers are determined by gas chromatograph/high-resolution mass spectrometry (GC/HRMS) while correcting the recovery rates. All of them are added up and the total is defined as the "PCB total concentration". Separate from the above, a standard is provided for CB28, CB52, CB101, CB118, CB138, CB153 and CB180, and the respective quantitative values are obtained. For DDTs and their dericatives, (i.e. o,p'-DDT, p,p'-DDT, o,p'-DDE, p,p'- DDE, o,p'- DDD, p,p'- DDD), a ^{13}C labeled surrogate standard substance is added to each of them so that their concentrations are determined while correcting the recovery rates. Then, all of them are added up and the total is defined as the "DDTs total concentration".

For chlordanes, a ^{13}C labeled surrogate standard substance is added to five types of chlordanes (i.e. trans-chlordane, cis-chlordane, trans-nonachlor, cis-nonachlor and oxychlordane) so that their concentrations are determined while correcting the recovery rates. Then all of them are added up and the total is defined as the "chlordanes total concentration".

For the other five substances, i.e. aldrin, dieldrin, endrin, heptachlor (including heptachlor epoxide) and HCB, a ^{13}C labeled compound is added to them as a surrogate standard substance, and their determinations are carried out while correcting the recovery rates in the pretreatment process.

In principle, domestic production and importation of the substances mentioned above are already prohibited. The results of investigation of Chemicals in the Environment reveal that the current concentrations of those substances in the environment are very low generally. Therefore, it is necessary to employ detection methods with particularly high sensitivity and selectivity for determination. The determination of any of the substances should be based on the use of high-resolution capillary gas chromatography, and high-resolution SIM measurement should be performed by EI ionization using a sector type large-scale mass spectrometer and setting the mass resolution capacity at 10,000 or above. In case that there is no problem regarding sensitivity and selectivity, a combination of compact Quadrapole type MS and NCI (negative chemical ionization) or ECNI (electronic capture negative ionization) may be used.

ANALYSIS OF PERSISTENT ORGANIC POLLUTANTS IN HUMAN BLOOD

Declining concentrations of persistent organic pollutants (POPs) in the environment and human tissues are evidence of the effectiveness of the Stockholm Convention. Human blood is one biological indicator recommended for monitoring trends of POP concentrations in the general population. Many of the validated analytical procedures used for analysis of POPs in human blood are chosen with respect to the analytical capacity of the specific laboratory concerned. Laboratory infrastructure, instrumentation, experience in trace analysis, and the ability to meet accompanying operational costs are examples of circumstances where a laboratorys choice and combination of sample preparation and detection method may vary. Yet, the analytical procedures adapted under these conditions may be specific and sensitive enough, even in laboratories with limited resources, to monitor concentrations of the Stockholm Convention POPs. Moreover, there are laboratories able to go even further and advance their analytical procedures by reaching detection levels lower than the well-established "state of the art" ultratrace levels or parts-per-quadrillion (ppq) levels. Such sub-ultra-trace level (or sub-ppq) procedures are, despite their great analytical capacity, limited to specialized laboratories having high-performance instrumentation and long experience in POPs analysis. However, in the future, if the effectiveness of the Stockholm Convention meets the expectations of continuously declining concentrations in human blood, what demands does that put on analytical performance? Will there be a need for today's emerging procedures for sub-ultra-trace levels?

In addition to laboratory conditions and instrumental performance, there are other important analytical aspects (e.g., those targeted by epidemiological studies). Epidemiological studies, in general, require a large number of samples and a careful selection of relevant compounds to be able to study the associations between compounds of toxicological relevance, or mixtures of these, and health effects in selected populations. What are the current trends in analytical procedures used in large monitoring and epidemiological studies?

Methods for Analysis of POPs in Human Blood

A number of procedures can be used for analysis of the chlorinated, brominated, and fluorinated classes of POPs in human blood, and the development and application of such procedures have been advanced by several authors pointing out a number of important factors regarding sampling, sample preparation, and instrumental analysis. The following two sub-sections provide summaries of analytical methods used for analysis of chlorinated or brominated compounds, and fluorinated compounds included in the Stockholm Convention. Sample-preparation procedures for human blood have been developed to extract several classes of compounds simultaneously when the analytes of interest share similar physico-chemical properties. The chlorinated and brominated compounds (e.g., PCBs, OC pesticides, and PBDEs) are neutral and lipophilic. By contrast, fluorinated compounds [e.g., the perfluoroalkyl acids (PFAAs)] are amphipathic and bind to proteins. As a consequence, different analytical strategies are applied to the group of chlorinated and brominated compounds compared to the fluorinated. Several extraction methods have been described including cavity dispersed microwave-assisted extraction (MAE), focused MAE, and pressurized fluid extraction (PFE), but liquid-liquid extraction (LLE) and solid-phase extraction (SPE) remain the most frequently used.

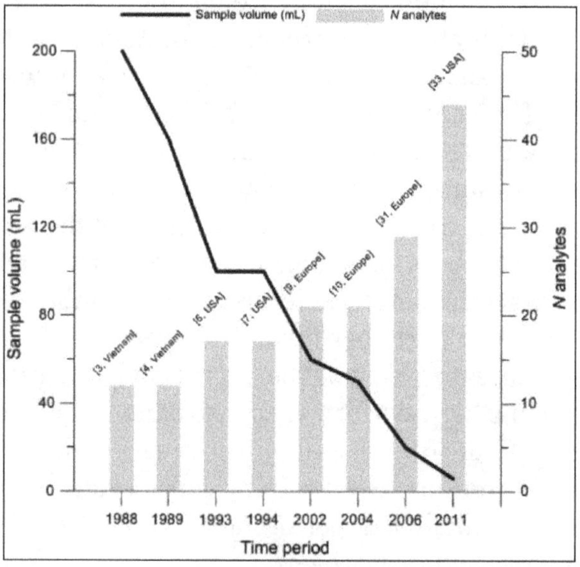

Over 30 years of analytical improvements in the analysis of dioxin and furans has resulted in decreased sample volumes and increased number (N) of analytes, including other classes of POPs. The number of target compounds included in sample-preparation, separation and detection methods has increased, and it is currently possible to monitor up to six compound classes simultaneously using small amounts of blood. The origins of the blood sample and the reference number cited are included in brackets.

In general, the analytical procedure requires 0.5–40 mL of blood sample, depending on the compounds of interest, followed by SPE or LLE. The instrumental analysis of PCBs, OC pesticides, and brominated flame retardants (BFRs) (e.g., PBDEs) is performed using GC often coupled to low-resolution MS instruments (GC-LRMS) while the analysis of chlorinated and brominated dioxins and dioxin-like (dl) PCBs generally requires larger amounts of sample and is performed using GC-HRMS. Analysis of the PFAAs [e.g., perfluoroalkyl carboxylic acids (PFCAs) and perfluoroalkane sulfonic acids (PFSAs)] typically involves SPE, LLE, or protein precipitation (PP). The instrumental analysis is usually performed using LC coupled to tandem MS instruments (LC-MS/MS). In general, it seems that the analysis in various monitoring and epidemiological studies moves towards more automated systems and comprehensive procedures using smaller amounts of blood. Correspondingly, the trend also moves towards more use of analytical procedures in which the different chlorinated and brominated classes of Stockholm Convention POPs are analyzed simultaneously.

Chlorinated and Brominated Pops

Pre-treatment

Denaturation and precipitation of blood proteins increases the effectiveness of extraction by releasing the lipophilic analytes from the sample matrix. Formic acid and solvents (e.g., n-propanol, and iso-propanol) are the most commonly used today. Moreover, formic acid provides enhanced extraction efficiency and is most commonly used for denaturation of proteins in human-blood analysis.

Extraction

LLE was previously widely used for extraction of POPs from human blood. Today, however, the use of LLE is not as frequent, and LLE methods are considered more labor intensive than SPE methods.

SPE was introduced as an alternative to LLE and works by extracting the target analytes by liquid-solid chromatography. SPE offers many advantages over LLE, but being unable to perform gravimetrical lipid determination is considered a drawback of SPE. Blood lipids (e.g., cholesterol, triglycerides, and phospholipids) are partly retained on the sorbent phase and lipid determination has to be done separately using enzymatic or colorimetric methods. As the sample passes through sorbent particles, the targets analytes are retained at the same time as a large portion of the interferences are washed out.

One of the major advantages of using SPE is the ability to remove interferences by washing the cartridges prior to elution using a non-polar solvent. The choices of sorbent phase and solvent system are very important for an effective SPE procedure. In short, sorbent phases are divided into four different types depending on their properties. Normal phase, reversed phase, adsorption, and ion exchange have been tested and shown to be suitable for various target analytes. Current SPE methods used for analysis of PCBs, OC pesticides and PBDEs are usually based on reversed-phase alkyl-bonded silica particles. In addition, polymeric reversed phase (e.g., the hydrophilic-lipophilic-balanced reversed phase sorbents) are widely used for both chlorinated and brominated compounds and found highly efficient in extracting the PBDEs. Consequently, it seems that the various types and brands (e.g., Oasis HLB and StrataX) of reversed-phase sorbents are most suitable for analysis of both chlorinated and brominated POPs.

Many of the SPE methods described are efficient in isolating multiple compounds, but usually require 0.5–40 mL of sample, depending on the target analytes. The challenge of obtaining lower limits of detection (LODs) can of course be met by using larger sample amounts. However, solvent volumes and handling times will also increase with the sample amount. One major challenge of using low sample amounts relates to the background concentrations in the laboratory environment resulting in high blank contamination. Concentrations of PBDEs, OC pesticides, and PCBs in the laboratory environment can fluctuate but are often abundant if the laboratory is used for analysis of highly-contaminated materials. Because the blank contamination defines the LOD, the blank contamination also sets the limit of sample amount required and which analytes are detectable. Thus, control of the blank contamination from the laboratory environment is very important when establishing a procedure for analysis of POPs in human blood.

A number of manual extraction procedures for multicompound analyses of chlorinated and brominated compounds in small volumes of sample have been presented. One procedure uses 4 mL serum and SPE for extraction of a total of 59 compounds, including PBDEs, PCBs, polybrominated biphenyls (PBBs) and selected OC pesticides prior to instrumental analysis using two-dimensional GC with time-of-flight MS (GCxGCTOF-MS).

Multi-compound analytical procedures that require less than 2 mL of sample have also been reported. Figure shows an example of a 0.5 mL sample extracted by SPE using Oasis HLB and clean-up using acid silica followed by multi-compound GC-HRMS analysis.

Recently, automated and semi-automated SPE methods, able to provide better reproducibility, higher sample throughput, and less contamination due to limited sample handling compared to manual SPE, were developed for analysis of chlorinated and brominated compounds.

Clean-up

Procedures to eliminate interfering substances are necessary after extraction. The clean-up step eliminates lipids that co-elute with the analytes. In general, the clean-up step is determined by three factors. First, the amount of sample used, second the extraction technique employed, and finally the chemical properties of the interferences.

Clean-up procedures are often based on liquid-solid chromatography using activated silica, active alumina, and active carbon materials. One widely used approach to remove lipids and other interferences is based on mixed (neutral, acidified or basic) activated silica columns. The activated silica treatment is based on fine materials packed in a glass column, where the size of the column and material depend on the amount of sample used. This manual approach was developed in the late 1970s and is, with modifications, still commonly used today.

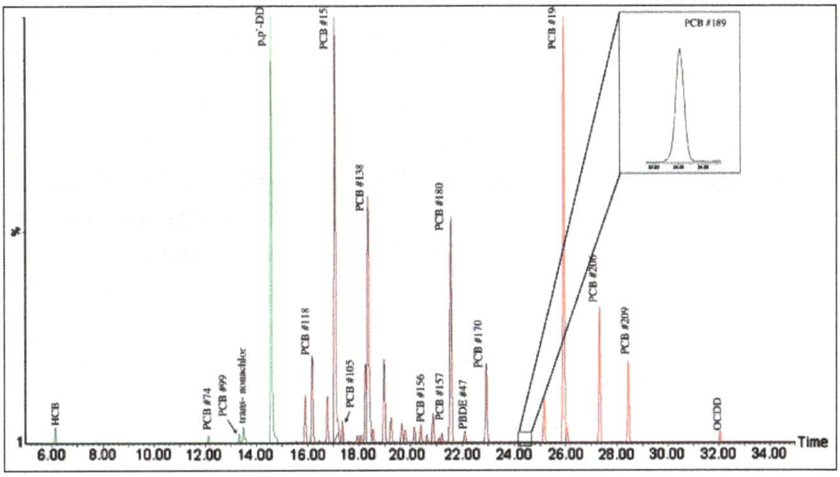

Multi-compound GC-HRMS analysis of selected OC pesticides and PCBs, including one PBDE and PCDD congener, using 0.5 Ml human plasma and SPE with multilayer silica clean-up.

Instrumental Analysis

Various injection, ionization, and detection techniques are used, depending on the target compounds and their concentrations in the samples. Injection is typically performed using the splitless injection mode, but large-volume injection (LVI) using programmable temperature vaporization (PTV) is increasing, as PTV has been shown to increase sensitivity compared to splitless injection-mode methods. Ionization can be performed using electron impact (EI), electron-capture negative ionization (ECNI), or electron-capture ionization (ECI). OC pesticides and PCBs can be analyzed using non-mass specific detection (e.g., GC-ECD) or mass-specific GC-LRMS operated in EI mode. However, detection using GC-ECD is not as robust for blood analysis as GC-LRMS methods and should eventually be phased out. But GCECD is still used for identifying populations with high exposures to OC pesticides and PCBs and for capacity building in analysis of POPs for the Stockholm Convention.

PBDEs are often analyzed using GC-LRMS operated in ECNI mode or by GC-HRMS operated in EI mode. Chlorinated and brominated dioxins and furans are analyzed using GC-HRMS operated in EI mode. Today, ultra-trace analysis of brominated and chlorinated POPs is performed using established GC-HRMS methods. Laboratories use GC-HRMS and GC-MS/MS for measuring almost all classes of chlorinated and brominated compounds including the polybrominated dioxins and furans (PBDDs and

PBDFs). Even though GC-HRMS provides the best selectivity and sensitivity for the large group of chlorinated and brominated compounds, there are other techniques able to provide sensitivity and selectivity similar to the GC-HRMS. For example, tandem GC coupled to TOF-MS (GCxGC-TOF-MS) showed good agreement with corresponding GC-HRMS analysis of selected chlorinated and brominated compounds. In addition, this technique offers some advantages over the GC-HRMS method, as it is possible to monitor 59 PCBs, PBBs, PBDEs, and OC pesticides by one single injection and at the same time identify co-eluting compounds having different fragmentation patterns by the use of mass-spectral deconvolution of overlapping peaks.

Cryogenic zone compression (CZC) is another technique that can be used in combination with comprehensive two-dimensional GC to enhance further the sensitivity of the GC-HRMS analysis. This technique combines GC with a cryogenic modulator with detection using HRMS operated in EI mode. This CZC-GC-HRMS configuration significantly enhances the signal of the analytes and makes it possible to detect down to ag levels of 2,3,7,8-TCDD and 1,2,3,7,8-PeCDD in a diluted 10 g serum extract without matrix effects. However, experiments performed on extracts from smaller serum samples suffered from high signals from the matrix, probably originating from the sample-preparation procedure.

Fluorinated POPs

Pre-treatment

A number of denaturants have been used for sample pre-treatment in the analysis of PFAAs [e.g., perfluorooctanoic acid (PFOA) and perfluorooctane sulfonic acid (PFOS)]. Studies show that formic acid, acetonitrile, and trichloroacetic acid, followed by vortex and centrifugation are the most effective denaturants for precipitation of proteins and red blood cells.

Extraction

SPE and LLE are the most commonly used extraction procedures for analysis of PFOS and other PFCAs and PFASs. Sorbent phases based on ion exchange and reversed phase are effective in extracting various PFASs from blood. Ion-exchange extraction uses attraction towards the negatively charged functional groups of the PFAAs to retain the analytes to sorbent phase while the reversed phase (or silica-based C18 sorbent) retains the analytes by the functional groups on the silica surface of the sorbent. Factors influencing the efficiency of the ion-exchange SPE are pH and solvent system. The pH of the sample needs to be one at which the compounds of interest are charged, and elution should be performed using a more polar solvent (e.g., methanol or acetonitrile). The principles of the manual SPE procedure were also applied to several recent automated methods.

In general, there are two possible ways to automate the SPE procedure when coupled to an LC-MS/MS system. The first fully-automated extraction, off-line SPE, performs the extraction by using miniaturized SPE cartridges for every sample injected.

The second automated extraction, on-line SPE (or column-switching), works by trapping the injected analytes onto an on-line SPE capillary column installed prior to the analytical capillary column. When the analytes have been trapped and focused onto the on-line SPE column, the system switches valves so that the analytes are eluted onto the analytical column with the mobile phase. After the cycle is complete, the on-line SPE is switched back from the analytical column and the on-line SPE is washed prior to the injection of the next sample. The main advantage of the on-line SPE (or column-switching) compared to the off-line SPE lies in the re-utilization of the on-line SPE. In general, automated SPE provides good reproducibility and low sample contamination due to limited sample treatment.

Clean-up

There are at least two different strategies to remove interferences from the sample extract. The first combines dispersed graphitized carbon (e.g., Envi-Carb) with glacial acetic acid, and the second removes interferences by filtration and provides less interference in the subsequent LC-MS/MS analysis. It is important to ensure that the membrane filters used are made of materials that are free from fluorinated compounds. Clean-up prior to instrumental analysis can also be performed by sample centrifugation.

Instrumental Analysis

The instrumental analysis of fluorinated compounds is preferably performed using HPLC-MS/MS operated in negative ESI mode. The majority of methods for analysis PFCAs and PFSAs in human blood use HPLC-MS/MS configurations. Ionization is mainly performed using ESI or turbo-ion spray (TIS), which is a type of ESI. Some PFASs can be derivatized and analyzed using GC-MS and thereby avoid PFOA contamination originating from the Teflon tubing in the LC system. However, such blank contamination problems can also be avoided by using a guard column prior to the injector.

Another major challenge in analysis of PFAS is the matrix effect that may enhance or suppress the detector signal and thus the recovery. In this context, using an isotope-labeled standard for each analyte reduces the matrix effect and variability in recoveries, and increases the accuracy of the MS quantification. In addition, matrix-matched calibrations can be used as an alternative to ID because the matrix-matched calibration curve, when compared to a calibration curve in a solvent, can compensate for matrix effects, and so reduce quantification errors from matrix effects. Finally, there are many considerations to accurate analysis of the PFAAs, including the removal of interferences prior to analysis, chromatographic separation of interferences on the LC, and finally selection of interference-free transitions in the MS/MS run.

PRESSURIZED LIQUID EXTRACTION AND AUTOMATED COLUMN CHROMATOGRAPHY SAMPLE CLEANUP

The Stockholm Convention on Persistent Organic Pollutants, which has been ratified by most countries in the world, aims at the elimination of persistent organic pollutants (POPs). The convention went into force in 2004. The continued worldwide interest in the compound classes covered by the convention has led to ongoing sample extraction, cleanup, and analysis in many environmental and food laboratories.

Originally, work focused on polychlorinated dibenzo-p-dioxins (PCDDs), furans (PCDFs), and polychlorinated biphenyls (PCBs), but other compound classes such as polycyclic aromatic hydrocarbons (PAHs) and polybrominated diphenylethers (PBDEs) have more recently drawn attention.

PCDDs and PCDFs are unwanted by-products of various processes such as combustion, incineration, metallurgical industry, and pulp and paper bleaching or production. A low natural background from, among other sources, forest fires was suggested in the 1970s. PAHs have some sources in common with PCDD and PCDFs, such as internal combustion and incineration. They also occur naturally. PCBs were intentionally produced from the 1920s to the 1970s for use in capacitators and transformers, as flame retardants, hydraulic fluids, sealants, and vacuum pump fluids. Total production has been estimated worldwide at 1.5 million metric tons. Similarly, PBDEs have been used and continue to be used as flame retardants around the world.

Although the levels of these compounds have been dropping in more recent years, they continue to be present in the environment and in foods at concentrations significant enough to pose threats to human health. These threats include possible damage to the immune system, the nervous system, and reproductive functions. In addition, they have been identified as endocrine disruptors, carcinogenic, and causing chloracne at higher levels of exposure.

These compounds also have low solubility in water and are chemically inert, which is the main reason that they resist environmental degradation. They can be found in many compartments of the environment, such as air, sediment, soil, and water. For environmental samples, the United States Environmental Protection Agency (US EPA) methods such as 1613 (PCDD/Fs), 1614 (PBDEs), and 1668 (PCBs) are generally followed. Because of low concentrations generally found on a per gram of sample basis (often in the nanogram to femtogram range) highly sensitive equipment, such as triple-quadrupole gas chromatography–mass spectrometry (GC–MS) and high-resolution GC–MS, is used for analysis. This analysis requires ultraclean sample processing (extraction, cleanup, concentration) to remove as many interferences as possible.

Experimental

Traditional techniques for sample extraction and cleanup have included Soxhlet extraction and open-column chromatography using various sorbents, such as silica and alumina. Soxhlet extraction can be a lengthy procedure with total extraction times up to 24 to 36 h. Typically an open system is used so this method of extraction can be prone to background contamination. The use of individual Soxhlet heating elements is also expensive with regards to the consumption of electrical power.

As an alternative, automated methods such as pressurized liquid extraction (PLE) have been developed. Any type of solid material can be extracted with this system. A solid or semisolid sample is weighed, mixed with inert drying material (such as Hydromatrix or Ottawa Sand) and placed in a stainless steel extraction cell capped on both ends with PTFE disposable caps. Porous metal frits in the end caps prevent clogging. Various cell sizes of 5–250 mL are used. Samples are usually spiked at this point with relevant internal or isotope dilution standards. The capped cell is placed in the position on the extraction unit and filled from bottom to top with extraction solvent. This solvent is typically 50:50 (v/v) dichloromethane–hexane but other mixtures or solvents can be used. The cell is pressurized at 1500 psi and heated at 120 °C for about 20 min, cooled for 15 min using a cooling fan, and flushed with extraction solvent (trapped in collection tube) and nitrogen (also in collection tube) to purge out any excess extraction solvent left. The whole procedure takes approximately 45–50 min. All extraction data (pressure and temperature) are computer controlled using real-time software and all events are recorded for traceability. Plots of the parameters are available for automated documentation and data reporting.

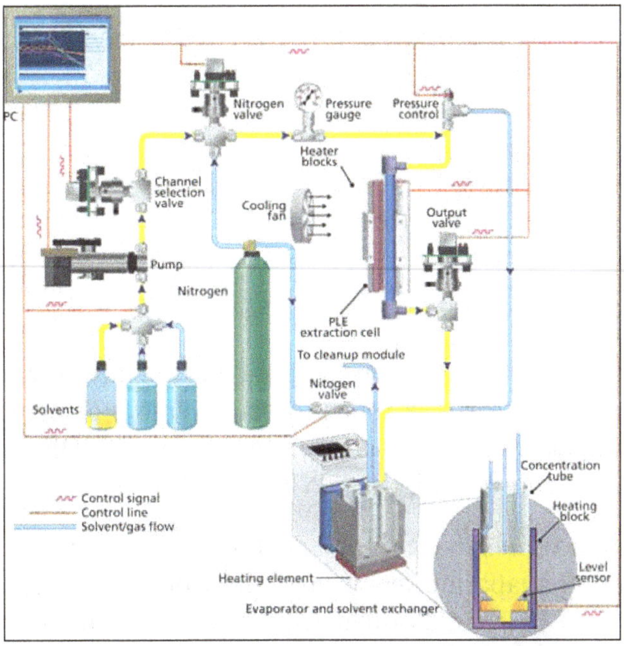

Schematic of a pressurized liquid extraction system.

Samples are reduced in volume in a six-position evaporator, which preheats them for 20 min at 60 °C, followed by heating under nitrogen at ~8 psi. The evaporator has built-in sensors that shut off the nitrogen flow when the sample reaches ~0.5 mL of volume. Since the subsequent cleanup requires the extracts to be free of dichloromethane, the samples are solvent exchanged to hexane during the volume reduction.

Further sample processing is accomplished by using automated column chromatography employing various certified prepackaged columns (silica, carbon, and alumina). The system consists of a control module, valve modules, pump modules, and sample processing modules. A small computer has been included in the system that controls the elution programs. Programming of the various columns, solvent flows, and volumes is done via a plumbing schematic. The system uses flows typically between 5 and 10 mL/min. The columns used are acid-base-neutral silica, carbon, and alumina. This is also the order in which samples are eluted through the system, in contrast to the more typical order silica-alumina-carbon. The silica column has different types of lipid capacity depending on the amount of silica used in the column. The mini version is capable of handling samples with up to 0.2 g of lipid, whereas the high capacity one is able to handle up to 5 g of lipid.

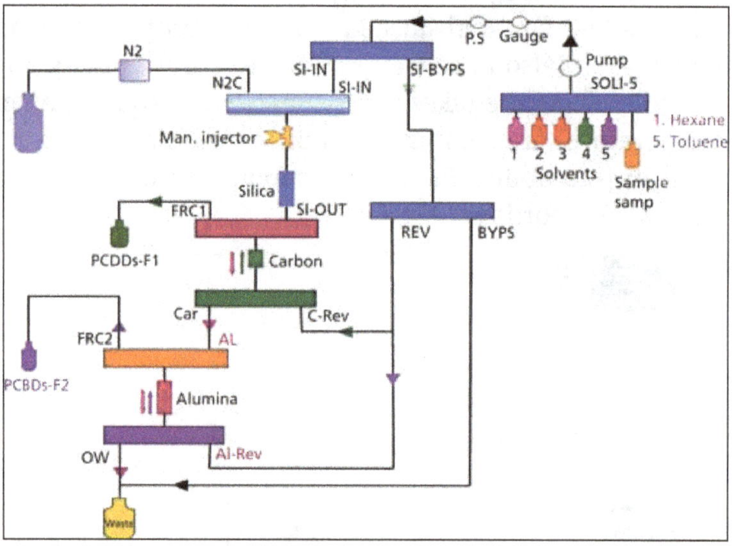

Schematic of a system for automated column chromatography cleanup.

The acid-base-neutral silica will remove components and interferences that can be oxidized or reduced. Alumina serves, among others, to remove chlorinated benzenes. The carbon column retains flat molecules such as PCDDs and PCDFs and co-planar PCBs while letting through mono- and di-ortho PCBs and PBDEs.

The elution program uses hexane and toluene, but no dichloromethane. This is an important consideration as laboratories around the world are trying to phase out this solvent. Columns were first conditioned with hexane. Then the various samples in hexane were loaded automatically onto the system. The silica column was eluted with hexane to transfer the analytes (PCDDs, PCDFs, PCBs) onto the carbon and alumina columns. The volume of hexane used was such that mono- and di-ortho PCBs were passing through the carbon column onto the alumina column. The carbon column was eluted with toluene collecting PCDDs and PCDFs and co-planar PCBs as fraction 1. The alumina column was eluted in the last step with toluene collecting the mono- and di-ortho PCBs as fraction 2. When PBDEs are the desired analytes for collection, these are collected with the PCB fraction in fraction 2. It should be pointed out that the cleanup described uses low volumes of solvent with the totals of hexane and toluene combined between 100–250 mL, depending on which type of silica column is used.

After collection, the sample volume was reduced to 1 mL using the evaporation step. Samples were then transferred to GC vials and ^{13}C-labeled recovery standard was added. Direct-to-vial connects, which can be screwed onto the collection tubes at one side and onto GC vials at the other side, offer an alternative approach in which sample transfer is not necessary. After the sample is in the GC vial, final volume reduction (typically to 10 μL) is achieved in a 24-position vial evaporator under a gentle stream of nitrogen at ambient temperature.

Analysis is done via splitless injection of 1–2 μL of sample onto the GC column, which is coupled with a Thermo single-quadrupole or triple-quadrupole mass spectrometer (PBDEs) or a Thermo high-resolution magnetic sector instrument (PCDDs, PCDFs, and PCBs). Typically, two ions are monitored for each compound analyzed (multiple ion detection). The temperature programs used usually have starting temperatures at 90–130 °C followed by a relatively steep temperature increase in the beginning and more gradual programming later in the run. PAHs were analyzed on a Thermo Scientific Trace GC system with a DSQ mass spectrometer.

ANALYSIS OF REMOTE ZONES AIR QUALITY

Remote zones represent the preferential observatory for understanding the evolution of our Planet and global change processes. Unfortunately, there is no disguising the fact that researches in these areas can represent a challenge in scientific terms. In addition, geographic distance and wildness of sites implies heavy capital expenditure by research units and involvement of highly trained staff able to realize surveys even in case of unexpected difficulties.

The remotest zone of the Earth is probably the Antarctic continent. Atlantic Pacific and Indian Oceans surround this continent; Antarctica is about 1000, 3800 and 2500 km from South America, Africa and Australia, respectively. Antarctica, on average, is the coldest, driest, and windiest continent, and has the highest average elevation of all the continents. Antarctica can be considered as a "frozen desert" with little precipitation; the South Pole itself receives less than 10 cm per year, on average. Temperatures reach a minimum of −80/−90 °C in the interior during winter and reach a maximum of 5/15 °C near the coast during summer. For a long time its climate, remoteness and physical geography discouraged exploration of the continent.

Remote Zones Contamination

In virtue of Antarctic climatic and geographic features, remote zones contamination studies are referred above all to Antarctica; insomuch as, Antarctica was designated by United Nations Environment Programme (UNEP) as a region for the global assessment of persistent toxic substances.

As about 98% of the continent is permanently covered by ice, Antarctica, the remotest continent in the Southern Hemisphere, is usually perceived as a symbol of the last great wilderness untouched by human disturbance. Unfortunately, the Antarctic environment is no longer pristine, even Antarctica is affected by global anthropogenic activities notwithstanding the peculiar characteristics of this continent. Indeed, in Antarctica there is no industrial activity or agriculture. Antarctica has a negligible non-native population, human presence is concerned largely with scientific investigations and related logistics operations (about 1,000 in winter to about 5,000 in the summer). Natural "barriers" such as oceanic and atmospheric circulation protect Antarctica from lower latitude water and air masses characterize Antarctica. Notwithstanding persistent organic pollutants were found in air, snow, mosses, lichens and marine organisms, showing that these contaminants are transported from other continents in the Southern Hemisphere. To give an example, pesticides have been neither produced or applied in Antarctica, but since 1960s several pesticides have been reported in biotic and abiotic matrices from Antarctica and the Southern Ocean.

POP Sources in Remote Zones

The Antarctic region in characterised by the almost total absence of local pollution sources. Pollutants reach Antarctica almost exclusively by long-range transport processes. Source identification of POPs and input sources in the Antarctic area is a very important and complex task.

As POPs released to the environment have been shown to travel vast distances from their original source, they can reach remote zones. For instance, since in Antarctica there are no agriculture or insects, chemicals as organochlorine pesticides are not used, their presence in Antarctic ecosystem can be due almost exclusively to long-range atmospheric transport from areas where these substances are still used and produced.

The main sources of POPs in the Southern Hemisphere are urbanized areas characterized by intensive agriculture, and tropical and subtropical regions, in particular South America, where pesticides are largely used. Another source of POPs is e.g. PCBs used in older electrical devices that were deposited as landfill in some developing countries. In addition, population growth and industrial development in several countries of the Southern Hemisphere are changing the global pattern of persistent anthropogenic contaminants and new classes of chemicals have already been detected in the Antarctic environment.

Locally, the greatest impact can be expected where research is carried out at long-term stations with a slight emission of some POPs. In addition, pelagic fisheries are an important operation in some parts of the region, and tourism is an increasingly important activity. Unlike pesticides, some industrial POPs such as PCBs were used in Antarctica leading to contamination in neighbouring areas, e.g. at McMurdo Station (Ross Island). The impact of scientific bases and related activities on the environment are more efficiently recorded by sediment analysis.

Air Analysis of Pops

The sampling and the analysis of persistent organic pollutants in atmosphere remain a challenge, especially when the atmosphere of remote zone is considered, taking into account low POP atmospheric concentrations.

Undoubtedly, the most crucial step is the sampling step. The gathered sample has to be representative of the monitored atmosphere; the sample has to represent an aliquot of the monitored atmosphere with the same composition and characteristics. Furthermore, from sampling to analysis time no sample contamination (analyte addition), degradation (analyte transformation) or loss (analyte loss), have to occur. Thus, sampling devices have to be appropriate even according to storage and transport conditions.

When a gas-phase component of the atmosphere has to be measured, two main approaches are operable during the sampling step: air sampling and analyte sampling. Air sampling consists in collecting an aliquot of atmosphere in special containers, the whole-air samplers, as plastic bags, glass vessels and stainless-steel containers. Analyte sampling consists in (passive or active) collecting only the gaseous analyte in a suitable "trap" avoiding air gathering. The approach's choice depends on several factors as chemical and physical characteristics of the air contaminant, its expected concentration, presence of interfering substances and ambient temperature and humidity. The selection of the sampling method depends also on analytical method detection limit: prior to sampling the minimum sample volume allowed by the analytical method has to be determined. Collection of the minimum volume is necessary to collect enough contaminant to meet the method's detection limit and thus measuring the expected concentration level. Therefore when the minimum volume exceeds e.g. 10 L, whole air sampling approach is unfeasible (the volume of whole air samplers is limited, in addition such an approach would be uneconomical) and thus analyte sampling approach has to be followed. This last is the case with semi-volatile compounds characterized by very low atmospheric levels, as PCBs.

Due to the low concentrations of POPs in atmosphere, an enrichment step is always required, i.e. a direct analysis of the air sample is unworkable. This step takes place just prior to analysis when a whole-air sample is collected, or at the same time of analyte sampling when the analyte is collected and then concentrated into the "trap".

The enrichment or trapping techniques can be based on three principles that "entrap" chemicals: absorption, adsorption and condensation. The analytes can be: a) collected into a solvent or on impregnated surfaces; b) collected onto adsorbents via passive or active sampling or onto fibres; c) condensed in cryotraps. The choice of the enrichment techniques strongly depends on the sample and on the contaminants to be collected.

Once decided the sampling method in accordance to needed sample type (instantaneous or time-integrated sample), sample number and above all expected chemical concentration and detection limit, the analytical question has to be faced. Compared to other chemicals and matrices, the analysis of POPs in atmosphere is time-consuming, complex, difficult, and expensive. Extremely sensitive, specific and efficient analytical procedures are required because the analysis must be performed for very low concentrations in the pg m^{-3} and sub pg m^{-3} range. Depending on the case, the pre-treatment step will consist in enrichment and desorption of the chemicals (in case of air sampling) or in extraction of the chemicals from the "traps" (in case of analyte sampling).

In consideration of chemical nature of chosen substances, analytical separation is generally carried out by high-resolution gas chromatography (GC). The detection can be realized by flame ionization detector (FID) or sometimes by electron capture detector (ECD), for electron-absorbing components such as halogenated compounds. However usually, detection and thus a certain identification of chemicals, is performed by high resolution mass spectrometer (MS) operated in electron impact ionization mode and in selected ion monitoring (SIM) mode.

POP Sampling and Analysis in the Atmosphere of Remote Zones

Air quality studies in remote zones require above all specific sampling and analytical methods that consider extreme environmental conditions. Sampling and analytical methods and equipment have to be specifically planned and modified in order to overcome several hurdles. Sampling devices have to own specific features able to realize representative gatherings even at very low temperatures, made up in material resistant to transport under adverse conditions and with relatively long-time storage capacity taking into account that several weeks might elapse between collection time and analysis. In addition, when sample collection has to be carried out at remote sites, the sampling equipment should be portable. On the other hand, analytical methods have to be at the cutting edge, as the low concentration levels of chemicals in the considered matrix; at times, a significant improvement concerning the current analytical methodology, in terms of precision and limits of detection can be required.

Whole-air samplings can be performed with a battery operated diaphragm pump and stainless steel canisters. As the pump is battery operated and oil-free, in situ samplings can be performed even in absence of electric current and no contamination of the air sample during the collection can occur. As the canisters are made in stainless steel with interior surface fully deactivated with SUMMA electro polish technique, any adsorption, deposition and release phenomena are avoided. In addition, air can be collected in them under pressure, consequently a greater volume of sample is available for performing several analyses and, at the same time, any contamination from external environment is thus avoided and on-line analyses can be carried out without a lift pump. The sampling apparatus described is durable, quite light, easy to transport and allows long-term storage. The commercially available apparatus is inappropriate for the severe Antarctic temperatures. Therefore, it has to be suitably modified: a) covering the pump with insulating materials (expanded polyurethane foam and a woollen cap) and then connecting its air way-in to a stainless steel probe; b) realizing a self-warmed and insulated warming system to warm up the pump and compensate the battery charge loss during the sampling operations and to store the pump inside it during transport to the sampling site, to minimize temperature effects; c) replacing canister's original connections and valves with 316 stainless steel connections with metal-to-metal seals and 316 stainless steel diaphragm valves in order to guarantee canister's hermetic seals even at extremely low temperatures and during sudden changes of temperature. The sampling system can be easily cleaned with evacuation/helium filling operations prior to air collection.

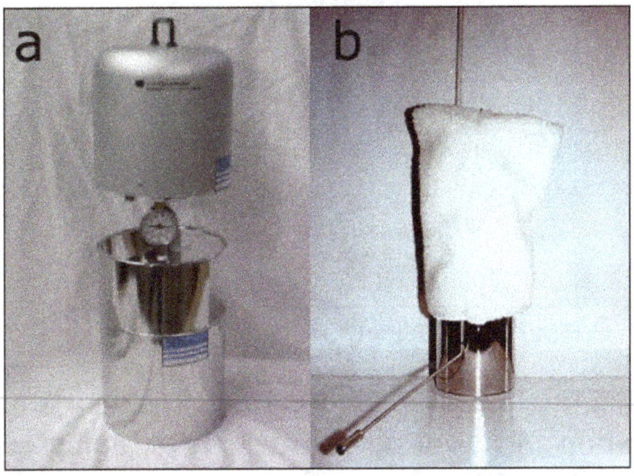

Air sampling equipment: Diaphragm pump and canister. (a) Original apparatus and (b) modified apparatus in usage conditions for polar climate with woollen blanket and stainless steel probe.

After whole-air sampling, a specific analytical methodology, characterized by high precision and accuracy and low limits of detection, has to be employed to quantify volatile organic compounds as VCHCs in air samples where the investigated substances can occur at pptv and sub-pptv levels. Volatile organic compounds can be analysed even in only slightly contaminated air samples, such as those from remote zones, with a specifically developed analytical technique based on a cryo focusing trap injector coupled to a gas chromatograph with mass spectrometric detection (CTI-GC-MS). The cryo focusing-trap-injector is the result of appropriate low cost modifications to an original purge-and-trap device to make it suitable for direct air analysis. Air samples, collected in stainless steel canisters, are introduced on-line directly into the CTI-GC-MS system; the volatiles contained in the air are cryo-concentrated into a cold trapat -120 °C and the entrapped volatiles are subsequently directly injected into the capillary column by means of rapid thermal desorption at 200 °C for the subsequent analysis of volatiles by GC-MS. Qualitative and quantitative analysis is performed by external standardization by means of certified gaseous standard mixture of the investigated VCHCs in synthetic air.

Active sampling of POPs from the atmosphere can be made with polyurethane foam plugs (PUF). Ambient air samples (from 1000 to 2500 m^3), collected over a period of 4-5 days using a high-volume sampling system, pass through the PUF and the compounds in the gas phase are adsorbed on the PUF. To be more precise, air is pulled through a glass-fibre filter to remove particles and subsequently passes through the PUF sorbent plugs to extract vapour phase organic compounds from the filtered air. The complete system is housed in a clear-anodized aluminium outdoor shelter.

Head of a PUF (polyurethane foam) sampler.

For POP determination, the PUF is Soxhelt-extracted with organic solvents for several hours. After concentration and clean-up, the extracts are analysed by gas chromatography-mass spectrometry (GC-MS) or by gas chromatography-electron capture detection (GC-ECD). In order to minimize the interfering substances contamination and to provide acceptable blanks, prior the sampling the PUF plugs have to be cleaned by a long Soxhlet extraction, then the cleaned PUF plugs are dried in a dessicator, packed in glass flasks and sealed in plastic bags until sampling. PUF plugs are convenient to handle and economical, but can exhibit breakthrough of semivolatile and volatile compounds.

Passive sampling of POPs from the atmosphere can be made with semipermeable membrane devices (SPMDs). These samplers are constituted by a thin film of triolein lipid enclosed in lay-flat, low-density polyethylene membrane. The polymer consists of transport corridors of less than 10 Å in diameter; these pores allow for the selective diffusion of hydrophobic organic chemicals, which are then sequestered in the lipid phase. SPMDs have to be deployed in appropriate screen to protect them from direct wet-deposition and sunlight.

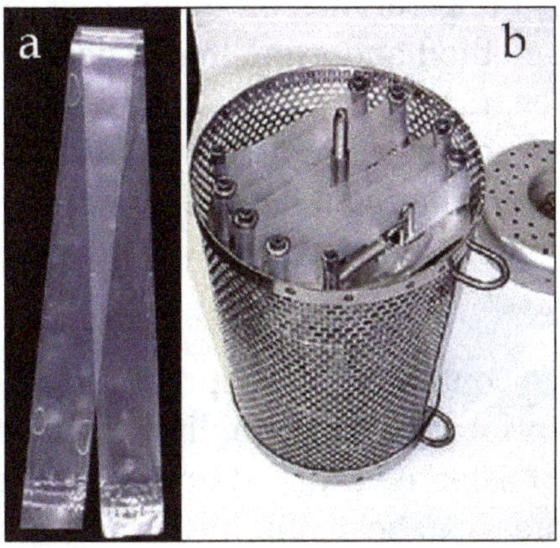

a) Semipermeable membrane device (SPMD); b) SPMDs deployed inside a stainless steel container.

For POP determination, SPMDs are dialysed with the appropriate solvent and the extract, after clean up, is analysed by GC-MS. SPMDs are deployed over weeks/months/years in the chosen site. Deployment time of the SPMD is analyte-dependent; in addition, it depends on instrument limit of detection, SPMD blank level, uptake rates and typical atmospheric concentration of the POP. The capacity of the SPMD depends on the quantity of lipid in the sampler and uptake rates by SPMDs depend on permeability of the membrane. Uptake rates are influenced by ambient temperature and wind speed; performance reference compounds can be used to "correct" for site-specific differences in uptake rates. SPMDs are easily contaminated, thus SPMD field blanks have to be taken to the field in sealed containers but not exposed, in order to evaluate the degree of contamination. Despite cited drawbacks, SPMDs are ideal samplers for studies into spatial differences in concentrations and profiles of POPs in the atmosphere, to identify sites where further investigations are required and for producing time-integrated data.

Quality Control

Quality control of the analytical procedure for POP determination in samples from remote zones is a very critical step in consideration of low atmospheric POP levels and the time lag between sampling and analysis. The risk of contamination is higher in particular for Antarctic air samples due to the much longer transport and storage time.

Suitable analytical quality control and quality assurance programs have to be run in the sampling site and in the laboratory, during sample gathering and analysis in order to guarantee data reliability.

With the purpose of assuring the overall quality and validity of experimental data, a systematic and detailed sampling strategy and blank choice have to be carried out. In order to prevent any contamination during sampling operations, clean room protocols should be attended. All the samplings should be performed downwind and away from local contamination sources, e.g. scientific stations. The operators should wear special clean room clothing and polyethylene gloves even during sampling, over their warm clothing. Immediately after the gathering, an opportune storage procedure (different according to the sampling method and sampling device) is applied in order to maintain unaltered the sample characteristics. The laboratories where the analysis will be performed, should be "Class 100 clean room", as classified by ISO Standard 14644-1. The chemicals used for sample treatment have to be ultra-clean reagents. The labware has to be extensively cleaned and conditioned with tested procedures to prevent any contamination.

Improvement of the storage containers to suppress the risk of contamination during storage should be realized prior to actual sampling. For instance, in order to evaluate the storage capability of a canister at very low temperatures and in case of sudden temperature changes, the canisters can be maintained at -20 °C (or at lower temperature if feasible) for a lapse of time comparable to Antarctic expedition duration and brought back to room temperature during the storage. Canister content, e.g. air with the lowest levels of organic contaminants, can be analysed immediately after the collection and after a variable storage time. No significant deviations of the composition during the storage should be noticed, at the investigated concentration level.

The risk of sample contamination has always to be investigated by taking field blanks. At least one field blank should be processed with each sampling episode. As an example, a field blank for POP sampling with PUF consists of a sample cartridge containing PUF and filter (spiked with a standard solution) shipped to the field, installed on the sampler and passively exposed at the sampling area. It is then sealed and returned to the laboratory for extraction, cleanup, and analysis, such as an actual sample.

A part from field blanks, procedural blanks have to be evaluated by analysing different types of blank samples, according to sampling and analytical procedures, e.g. laboratory method blanks and solvent blanks.

Finally, the samples processed have to give highly reproducible results by using traceable standards. Unfortunately, gas standard reference materials, with similar composition of remote zone atmosphere, are lacking.

Levels of POPs in Remote Zones Air

While measurements of POPs in the Northern Hemisphere are quite well documented, Southern Hemisphere measurements, and particularly POP measurements are still sparse and relatively limited. There is a limited atmospheric database on POPs in remote areas. Data about air levels of POPs are very sparse and originate mainly from cruises close to the Antarctic continent and from a few sampling campaigns on the continent itself.

The studies concerning the atmospheric content of POPs in Antarctica mostly focused PCBs, organo-chlorine pesticides as hexachlorocyclohexanes and dioxins and emerging pollutants such as VCHCs and PAHs, that were determined in atmospheric samples collected at different Antarctic sites. The studies carried out by several research groups all over the world, focused on long-range transport processes, latitudinal distribution of POPs, concentration temporal tends, scientific expedition impact and on future emerging pollutants.

Long-range Transport Studies

Measurements of POPs in the atmosphere of remote zones can be useful to estimate the contribution of long-range atmospheric transport to the input of different classes of POPs in the southern hemisphere. The results obtained by several research groups are consistent with the hypothesis of global fractionation and long-range transport.

The possibility of atmospheric long-range transport episodes from South America to Signy Island was pointed out by Kallen born in 1998 which measured POPs such as polychlorinated biphenyls and pesticides in ambient air at Signy Island (Antarctica), a sub-Antarctic sampling station, over a period of 17 weeks in the austral summer of 1994. Mean concentrations for single polychlorinated biphenyls (0.02-17 pg m^{-3}), for chlordanes (0.04-0.8 pg m^{-3}), DDT compounds (0.07-0.40 pg m^{-3}) and γ-hexachlorocyclohexane (22 pg m^{-3}), were comparable to those in Arctic air. Previous studies realized in by Tanabe et al., Weber & Montone and Larsson et al. provided further evidence of atmospheric transport of PCBs and organo-chlorine pesticides in the Antarctic environment. Tanabe et al. measured PCBs and chlorinated hydrocarbon pesticides such as DDTs and hexachlorocyclohexanes (HCHs) in air sampled in proximity of the Japanese research stations in Antarctica and adjacent oceans during December 1980 to March 1982.

The atmospheric concentrations of chlorinated hydrocarbons seemed to decrease in the transport process from northern lands to Antarctica, but the compositions of PCBs, DDT compounds and HCH isomers were relatively uniform throughout this process. Regional and seasonal variations were found in aerial concentrations of these pollutants at Syowa Station and adjacent seas in Antarctica. Larsson et al. monitored PCBs, p,p'-DDT, p,p'-DDE and γ-HCH in the lower atmosphere of Ross Island, in Antarctica for 2 yr. Geometrical means were 15.2 pg m^{-3} for PCBs, 2.0 pg m^{-3} for p,p'-DDT, 1.0 pg m^{-3} for p,p'-DDE and 25.8 pg m^{-3} for γ-HCH. Atmospheric levels of γ-HCH were positively correlated with temperature, and a significant difference was found between spring-summer and summer-winter concentrations. No season related differences were found for the other chlorinated hydrocarbons, possibly owing to their lower vapour pressure and the cold climate. Periods with increased atmospheric levels of PCBs and DDT compounds were recorded. Chlorinated pesticides in air were measured from the Greenpeace ship Gondwana between New Zealand and Ross Island, January-March, 1990. Geometric mean concentrations: α and γ-HCHs 4.0 and 3.8 pg m^{-3}, hexachlorobenzene

60 pg m^{-3}, heptachlor epoxide 0.52 pg m^{-3}, chlordanes and nonachlors 1.8 pg m^{-3}, p,p'-DDE and p,p'-DDT 0.81 pg m^{-3}. Concentrations of DDTs in Antarctic air were similar to those in the Arctic. A trend of decreasing pesticide concentration at higher latitudes was noted. Further evidence of atmospheric transport of PCBs in the Antarctic environment was provided by Fuoco et al. whose studies of Antarctic lake sediment found high contributions of atmospheric particulate matter and concluded that it is the primary vehicle of transport and diffusion of PCB in the environment. Montone et al. determined the atmosphere content of a large number of organo-chlorine compounds (HCB, HCHs, pp'-DDE, pp'-DDD, pp'-DDT and 11 PCB) in a survey performer in November of 1995 in the open ocean between 23 S and 62 S.

Air mass back trajectories were used to assess the origin of the air masses that arrived at the sampling point at low altitude. Organo-chlorine levels in the atmosphere ranged from not detected (<0.6) to 25.3 pg m^{-3} for HCB (with a mean concentration of 15 pg m^{-3}), 3.9–32.5 pg m^{-3} for ΣHCHs (α-HCH + γ-HCH) (with a mean concentration of 8.8 pg m^{-3} for Lindane), 3.7– 102.6 pg m^{-3} for PDDTs (pp'-DDT + pp'-DDD + pp'- DDE) and 46.2 up to 985.0 pg m^{-3} for PCBs. Choi et al. deployed passive air samplers for one year at the Korean polar research stations at Ny-Alesund, Norway and King George Island, Antarctica to assess levels and patterns of polychlorinated biphenyls (PCBs) and organo-chlorine pesticides in polar regions. The observed backward trajectories suggested that long-range transport from source regions in Northern Europe and Russia and the southern tip of South America, respectively affect the Arctic and Antarctic areas. The results were consistent with the hypothesis of global fractionation and long-range transport. The average level of Σ_{205}PCB (excluding three mono-CBs and PCB-11) was five times higher in the Arctic (95 pg m^{-3} range) than in the Antarctic (19 pg m^{-3}). The difference of concentration in terms of Σ_{205}PCBs reflects the hemispheric distribution of global PCB emissions. Levels of Σ_9PCB at Ny-Ålesund were similar to those reported for other Arctic sites, while levels at King George Island were lower than at other sites on the Antarctic Peninsula but 1 order of magnitude higher than background levels measured at a more remote Antarctic site. Light homologues were predominant in all samples, consistent with the hypothesis of global fractionation and predictions of long-range transport potential.

Latitudinal Distribution Studies

To establish POP background air concentrations in oceanic regions and to assess their latitudinal distribution, POP data concentration can be collected during cruises such as the Atlantic cruise from the United Kingdom to Halley, Antarctica.

With the purpose of give evidence of POP transport from the northern hemisphere to the southern and to reconstruct their trends in concentrations, Lakaschus and co-workers collected POP data during two cruises across the Atlantic Ocean between the Arctic Ocean (Ny-Alesund/Svalbard, 79 N; 12 E) and the Antarctic Continent (Neumayer Station/Ekstroem Ice Shelf, 70d S; 8.2 W) in 1999/2000. Hexachlorocyclohexanes

(HCHs) were determined simultaneously in air and seawater. The concentrations of α-HCH and γ-HCH in air and surface waters of the Arctic exceeded those in Antarctica by 1-2 orders of magnitude. Between Cape Town and Neumayer Station the average gaseous concentrations of α-HCH and γ-HCH were 1.1 pg m^{-3} and 1.4 pg m^{-3} and decreased further to 0.36 ± 0.03 pg m^{-3} and 0.15 ± 0.04 pg m^{-3} on the Ekstroem Ice Shelf at Neumayer Station. By comparison of results obtained from archived samples taken (by the same research group) in 1987/1997 and those from 1999, a systematic decrease of α-HCH was observed at all sampling locations. Booij and co-workers collected POP data during a cruise from the island of Texel (The Netherlands) to Walvis Bay (Namibia) and Cape Town (South Africa). Aerosol and surface water samples were collected and different classes of organic pollutants were determined. The content of PCBs, hexachlorobenzene (HCB), 1,1-dichloro-2,2-bis(4-chlorophenyl)ethene (4,4′-DDE), and polyaromatic hydrocarbons (PAHs) were determined. The air concentrations of HCB ranged from 56 to 145 pg m^{-3}, a value typical for background levels for air (4–140 pg m^{-3}). Atmospheric 4,4′-DDE concentrations ranged from 0.1 to 0.9 pg m-3 and were smaller than the values of observed by Iwata et al. in the same area during one cruise carried out in April 1990 (1.3–6.3 pg m^{-3}).

During the same cruise, the atmospheric content of polycyclic aromatic hydrocarbons (PAHs) and of one emerging class of pollutants (polychlorinated naphthalenes, PCNs) were determined. The highest PAH concentrations occurred in the European samples, and in samples close to West Africa and South Africa. Consistently low PAH concentrations were measured in the southern hemisphere open ocean samples (190-680 pg m^{-3}). Concentrations showed a diurnal cycle, the day/night ratios of phenanthrene, 1-methylphenanthrene and fluoranthene were typically 1.5-2.5:1. The highest PCN concentrations occurred in the European samples, but high values were also detected off the West African coast, and in the sample taken closest to South Africa.

Temporal Trend Studies

Studies regarding atmospheric temporal trends of POPs as organo-chlorine pesticides can be carried out by measuring the atmospheric content of HCB, heptachlor, α- and γ-HCH, heptachlor epoxide.

Dickhut et al. determined hexachlorobenzene (HCB), heptachlor, α- and γ- HCH and heptachlor epoxide in air, seawater, sea ice, and snow. Samples were collected during the austral winter and summer along a transect in the Western Antarctic Peninsula. By comparison with previous studies they concluded HCB and HCH levels decreased over the last twenty years, ΣHCH showed a half-life of 3 years in Antarctic air. On the other hand, heptachlor epoxide levels did not show such a reduction, possibly due to its continued use in the southern hemisphere. They detected peak heptachlor atmospheric concentrations in coincidence with air masses moving into the region from lower latitudes. Levels of γ-HCH were 1.2-200 times higher in annual sea ice and snow compared to α-HCH, likely due to greater atmospheric input of γ-HCH. Based on the ratio of α/γ-HCH <1 in Antarctic air, sea ice and snow they concluded that there is a predominance of influx of γ-HCH versus technical HCH to the regional environment. In addition, they also observed that the α/γ-HCH in seawater was >1, likely due to more rapid microbial degradation of γ- versus α-HCH. The water/air fugacity ratios for HCHs demonstrate continued atmospheric influx of HCHs to coastal Antarctic seas, particularly during late summer.

The spatial distribution of α-HCH and the net direction of air/water gas exchange were determined between November 1997 and February 1998 by Jantunen et al. Air and water samples were collected during a cruise between South Atlantic Ocean (South Africa) and Antarctica SANAE Base (70 S, 3 E). The α-HCH concentrations in air and surface water resulted much lower than in Arctic regions, consistent with the historically lower usage of technical HCH in the Southern Hemisphere. The water/air fugacity ratios of α-HCH were lower than or equal to 1.0, indicating steady state or net deposition conditions. The results of one analysis of the enantiomeric fractionation showed that the α-HCH in water was enantio selectively metabolized and that the two isomers [(-)α-HCH and (+)α-HCH] in the air boundary layer reflected those in surface water, showing the bidirectional nature of gas exchange.

Scientific Expedition and Station Impact Studies

Unlike pesticides, only some industrial POPs such as PCBs were used in Antarctica. Therefore, scientific expedition and bases can be a source of POPs.

In order to study the impact of scientific expeditions and bases, areas neighbouring stations and zones where occurred scientific operations, can be monitored. A concentration gradient was observed with distance from the main buildings of scientific bases in the Antarctic, which strongly indicated the influence of local sources.

ANALYSIS IN COMPLEX MATRICES

The analysis of persistent organic pollutants (POPs) in complex matrices is challenging because of their broad concentration ranges and the presence of interfering substances with similar physicochemical properties. In addition, there is a lack of instrument technology for widespread analyses of targeted and nontargeted compounds in minimally processed environmental samples. POPs in complex matrices such as sediment and fish samples were analyzed using gas chromatography (GC) with high-performance time-of-flight mass spectrometry (TOF-MS). High-performance TOF-MS offers the benefits of classical TOF (for example, a nonscanning mass analyzer, high acquisition rates, and spectral continuity) with the added advantages of increased mass accuracy and resolving power (R = 50,000 at FWHM in ultrahigh resolution mode). Different classes of POPs were screened in single sample analyses. The power of high-resolution

MS coupled with the ability to extract specific ions from full-range mass spectral data makes high-performance TOF-MS valuable for analyses of POPs in complex matrices.

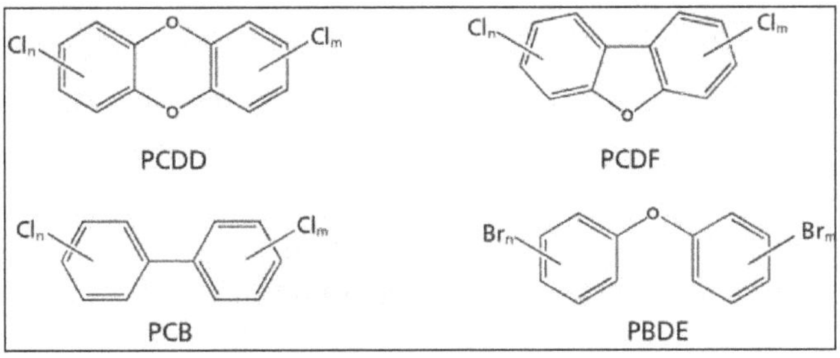

General structures for POPs.

Persistent organic pollutants (POPs) are halogenated organic compounds produced industrially for a wide variety of commercial applications. Examples of POPs include dioxins (polychlorinated dibenzodioxins [PCDDs] and polychlorinated dibenzofurans [PCDFs]), polychlorinated biphenyls (PCBs), and polybrominated diphenyl ethers (PBDEs). PCBs, marketed under trade names such as Aroclor, Sovol, and Clophen, were used for many years as dielectric fluids in electrical transformers and capacitors, heat transfer fluids, hydraulic fluids, lubricating and cutting oils, and additives in many commercial products. PBDEs have been used as fire retardants in household goods, construction material, and various plastics for the aviation and automotive industry. Dioxins are by-products formed during combustion processes. POPs are relatively inert and lipophilic, and tend to bioaccumulate in the environment. Their adverse health and environmental effects have led to their regulation in many industrialized countries. Detection of POPs is complicated by their broad range of concentrations and the presence of interfering substances with similar physicochemical properties. Furthermore, there has been a lack of appropriate instrumental technology for exhaustive analyses of targeted and nontargeted POPs in minimally processed environmental samples. Analysis of environmental samples typically requires costly equipment and highly specialized personnel. Analysts can spend significant time isolating, purifying, and characterizing the materials only to find that their efforts have resulted in partial loss, chemical modification, or complete elimination of targeted substances.

Comprehensive analysis of POPs should involve the study of all classes of POPs and their metabolites. Metabolites of POPs are emerging as regular contaminants in wildlife and are critical for the risk assessment of human exposure to pollutants such as PCBs. Detection of POPs can be improved by reducing the amount of sample processing or enhancing instrument selectivity. High-resolution mass spectrometry (MS) has been the method of choice for the analysis of POPs; however, time-of-flight (TOF) mass spectrometers have emerged as an attractive alternative to their more costly counterparts. TOF-MS instruments were commercially available in the 1950s, and recent advances such as orthogonal acceleration, extended path lengths, and improvements in data acquisition and processing have led to a renaissance of TOF mass analyzers as modern tools for biological and chemical research. The attractive features of TOF-MS, such as an unlimited mass range, fast acquisition rates, and spectral acquisition over a wide mass range without sacrifice of sensitivity, have been advantageous for the analysis of both small and large molecules. The fast acquisition rates of TOF-MS result in nonskewed mass spectral data that can be easily deconvoluted. Unfortunately, the low-resolution capabilities of classical TOF instruments made them unsuitable for the analysis of POPs. The emergence of high performance TOF-MS with extended flight paths, improved detection systems, and resolving powers greater than 50,000 (FWHM), has only increased the utility of these instruments for both targeted and nontargeted analysis of complex samples.

Experimental

Gas chromatography (GC)–TOF-MS analyses of fish and sediment samples containing PCBs and other POPs were performed using a Pegasus GC-HRT MS system (LECO Corporation, St. Joseph, Michigan) equipped with an Agilent Technologies (Palo Alto, California) 7890A GC system. PCB-containing sediment and fish samples were analyzed using an Rxi-5 Sil MS column (Restek, Bellefonte, Pennsylvania) and data were collected in high-resolution mode. Samples were previously prepared and not processed further before analysis. GC and MS parameters for PCB sample analyses are listed in table.

Table: GC–MS parameters for PCB sample analysis.

GC Conclusions	
column	60 m × 0.18 mm, 0.10- μm d_f Rxi-5 Sil MS (Restek)
Carrier	He, constant flow, 1.0 Ml/min
Inlet	275 °C
Oven	80 °C (1.0 min), ramped at 30 °C/min to 160 °C ramped at 3 °C/min to 300 °C
MS Conditions	
Transfer line	300 °C
Ionization	El at 74 ev
Mass range (m/z)	50-1000
Acquisition rate	2 or 3 spectra/s
Calibration	PFTBA

At the heart of the MS system is a folded flight path (FFP) mass analyzer with three operational modes: nominal (resolving power, R = 1000, FWHM at 219 m/z), high resolution (R = 25,000, FWHM at 218.985080), and ultrahigh resolution (R = 50,000, FWHM at 218.985080). Ions are introduced via orthogonal acceleration and then travel between a parallel set of gridless mirrors and through a set of periodic ion lenses.

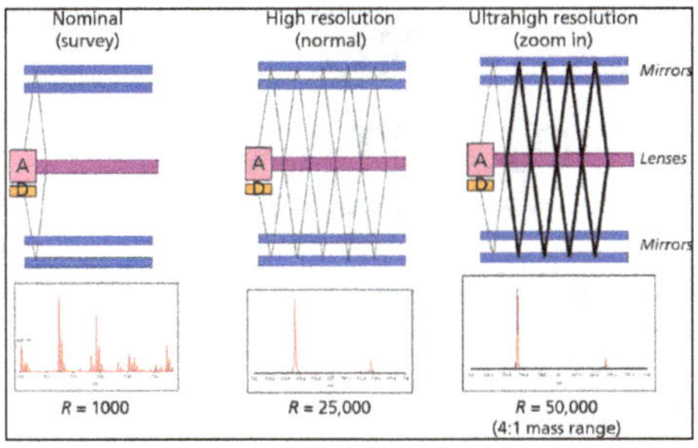

TOF-MS system operational modes (Pegasus GC-HRT). A = accelerator, D = detector.

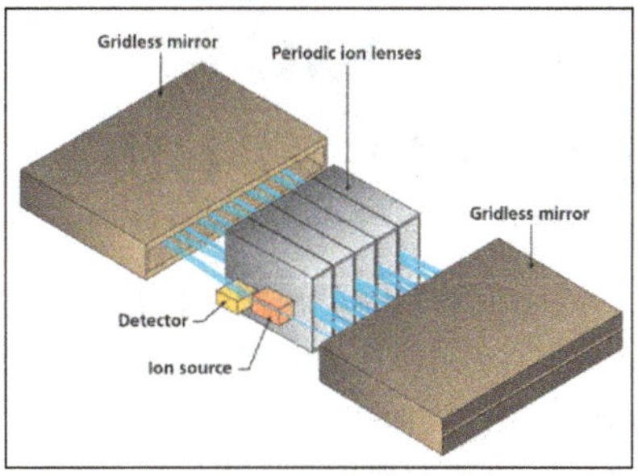

FFP mass analyzer.

The flight path can be adjusted to 2 m (nominal), 20 m (high resolution), or 40 m (ultrahigh resolution). Most routine analyses can be performed in high-resolution mode; however, nominal mode can be used for quick inspection of samples, chromatographic method development, or NIST library searches. Ultrahigh-resolution mode can be used for samples requiring a high level of selectivity.

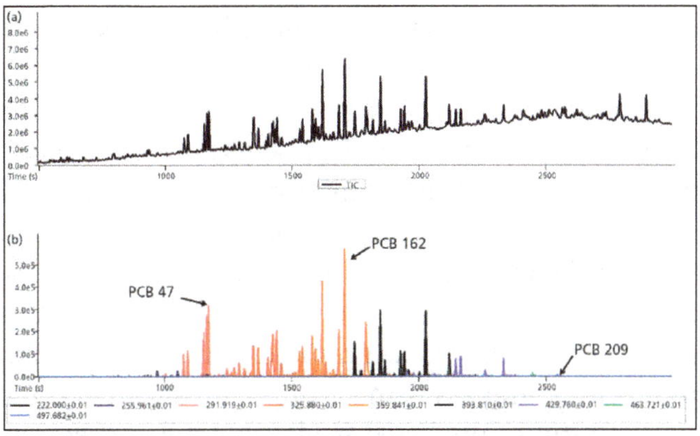

(a) TIC of sediment containing elevated PCB levels. (b) XIC of sediment sample showing three different PCBs.

Analysis of POPs is challenging due to the similar chemical and physical properties of these compounds. They are aromatic, lipophilic, and usually coextracted with numerous organic compounds when nonpolar solvents (for example, hexanes) are used during sample preparation. High-performance TOF-MS is capable of performing exact mass measurements and allows for selection and extraction of targeted analytes for elemental composition confirmation. High-performance TOF-MS provides a comprehensive profile of the sample in a single run, making it a powerful tool for the analysis of POPs in complex matrices.

Table: Isotope ratio values for p,p'-DDE, PCB 153, and PCB 209.

Analyte	Formula	R.T. (s)	Area	Ions	Calculated	Observed
p,p'-DDE	$C_{14}H_8Cl_4$	1482.06	56239	M+/M+2	0.776	0.784
PCB 153	$C_{12}H4Cl_6$	1531.479	739179	M+2/M+4	1.24	1.17
PCB 209	$C_{12}Cl_{10}$	2545.459	57494	M+2/M+4	0.693	0.705

MEASURING AND MONITORING PERSISTENT ORGANIC POLLUTANTS

Due to growing concerns regarding persistent organic pollutants (POPs) in the environment, extensive studies and monitoring programs have been carried out in the last two decades to determine their concentrations in water, sediment, and more recently, in biota. Whilst the vast majority of these efforts either attempt to compare (a) spatial changes (to identify "hot spots"), or (b) temporal changes to detect deterioration/improvement occurring in the environment, most studies could not provide sufficient statistical power to estimate concentrations of POPs in the environment and detect spatial and temporal changes. Despite various national POPs standards having been established, there has been a surprising paucity of emphasis in establishing accurate threshold concentrations that indicate potential significant threats to ecosystems and public health.

Although most monitoring programs attempt to check compliance through reference to certain "environmental quality objectives", it should be pointed out that many of these established standards are typically associated with a large degree of uncertainty and rely on a large number of assumptions, some of which may be arbitrary. Non-compliance should trigger concern, so that the problem can be tracked down and rectified, but non-compliance must not be interpreted in a simplistic and mechanical way. Contaminants occurring in the physical environment may not necessarily be biologically available, and even when they are bioavailable, they may not necessarily elicit adverse biological effects at the individual or population levels. As such, we here argue that routine monitoring and reporting of abiotic and biotic POPs concentrations could be of limited use, unless such data can be related directly to the assessment of public health and ecological risks. Risk can be inferred from the ratio of predicted environmental concentration (PEC) and the predicted no effect concentration (PNEC). Currently, the paucity of data does not allow accurate estimation of PNEC, and future endeavors should therefore, be devoted to determine the threshold concentrations of POPs that can cause undesirable biological effects on sensitive receivers and important biological components in the receiving environment (e.g. keystone species, populations with high energy flow values, etc.), to enable derivation of PNECs based on solid scientific evidence and reduce uncertainty. Using the threshold body burden of POPs required to elicit damages of lysosomal integrity in the green mussel (Perna virvidis) as an example, we illustrate how measurement of POPs in body tissue could be used in predicting environmental risk in a meaningful way.

TOTAL ORGANIC CARBON

The Total Organic Carbon (TOC) is one of the most important composite parameters in the assessment of the organic pollution of water. Since it includes all carbon compounds as one mass, it is exactly defined and an absolute quantity. Therefore, it may be determined directly.

- TC: "The organic and inorganic carbon in water, including the elemental carbon".
- TIC: "The carbon contained in water; elemental carbon, total carbon dioxide".
- TOC: "The organic carbon in water, bound on dissolved or suspended substances. Elemental carbon will be also determined".
- "The total organic carbon (TOC) is a measurement method for the content of carbon of dissolved and undissolved organic substances in water".

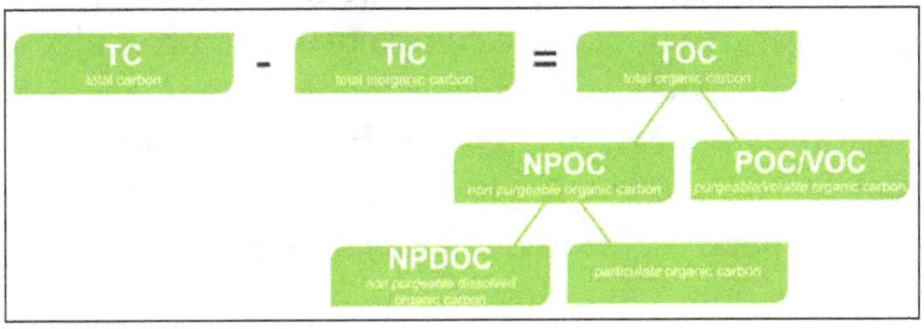

TOC Measurement Methods

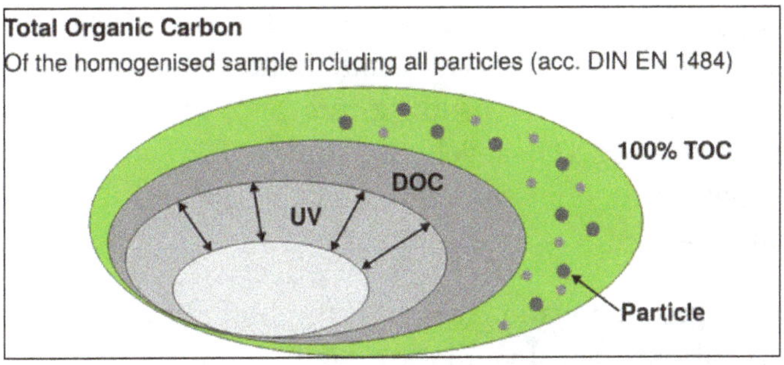

TOC contains dissolved and particulate organic carbon.

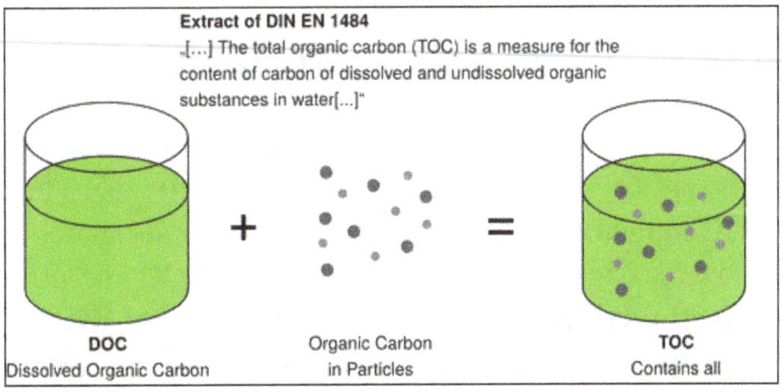

The difference between TOC and Some Organic Carbon.

In general, the TOC is determined by oxidizing a water sample. The produced CO_2 is detected and defined quantitatively. However, not all methods succeed in the complete oxidation of a sample. Often enough this may result only in SOC (Some Organic Carbon) instead of the TOC.

Thermal Oxidation

With this oxidation method the sample will be combusted in a reactor. Usually, a maximum temperature of about 1,000 °C will be reached, which however does not allow the complete oxidation of all carbon compounds. Therefore, a catalyst, such as copper oxide or platinum, must be used at this temperatures. Normally, the catalytic combustion method can handle a TOC concentration of up to 4,000 mg/l. To reach higher ranges the sample usually needs to be diluted with demin water.

LAR Process Analysers AG offers a unique and patented high temperature (HT) method at 1,200 °C. This temperature enables the complete oxidation of all carbon compounds without any catalysts. It measures TOC concentrations up to 50,000 mg/l without dilution. In a special heat resistant ceramic reactor the water sample is evaporated and all carbons are completely oxidized to CO_2 gas. Afterwards, the CO_2 concentration will be analysed with a Non-Dispersive Infrared (NDIR) detector. Thus, the TC, TOC and TIC can be determined within only 3 minutes.

This ultra high temperature method is used for both, the most challenging and highly contaminated waters ($\text{QuickTOC}_{\text{ultra}}$, $\text{QuickTOC}_{\text{airport}}$) and waters relativily free of solid maters ($\text{QuickTOC}_{\text{effluent}}$). LAR Process Analysers AG uses the batch injection method with the advantage that the analysers can easily handle sticky, oily and hard to oxidise dissolved and suspended organics resulting in a fast, reliable and accurate analysis. Even with rapidly fluctuating TOC levels the correct TOC concentration will be measured, whereby peaks throughout the course of the day are determined without any memory or adsorption effects.

For applications with purified water, this LAR method allows the patented simplified, ready-at-any-time calibration and validation method ($\text{QuickTOC}_{\text{condensate}}$, $\text{QuickTOC}_{\text{purity}}$, $\text{QuickTOC}_{\text{pharma}}$).

Photochemical Oxidation (UV-persulphate Method)

Here the TOC is oxidized by means of UV light and a digesting reagent, sodium persulphate, and the produced CO_2 is measured with a NDIR detector. This method suits the determination of TOC in clean water (drinking water, condensate, boiler feed water), since particles are hard to oxidize completely. The $\text{QuickTOC}_{\text{uv}}$ combines this technique with the direct TOC method or Non-Purgable Organic Carbon (NPOC) method, whereby the continuously provided water samples will be treated in a multi stage process.

Wet Chemical Oxidation

With this method the water sample is oxidized by means of strong chemicals as oxidants such as ozone, which are slightly dangerous to health and environment. The ozone oxidation acids and bases are used to adjust the pH value of the sample along the pH scale. However, the oxidation potential of such methods is relative, since particles and more complex carbon compounds can only be partially digested or not at all. With regards to the latest standards of occupational safety and environmental protection these methods are not recommended.

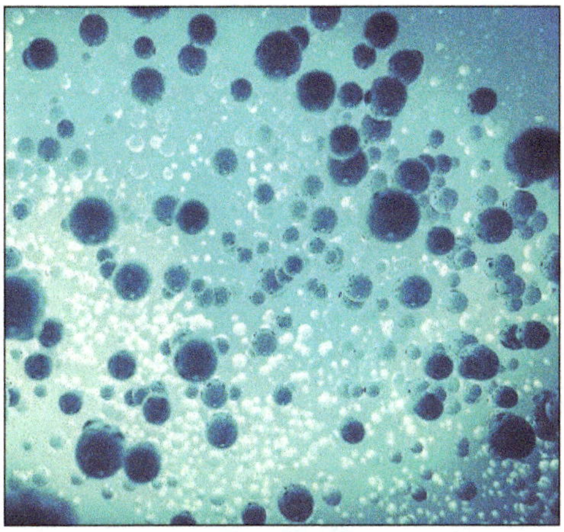

TOC is a valued, analytical technique that is applied by organizations and labs to determine how suitable a solution is for their processes. Unless it's ultrapure, water will naturally contain some organic compounds, understanding how much is key.

TOC has become an important parameter used to monitor overall levels of organic compounds present. This has happened despite the lack of any direct quantitative correlation between total organic carbon and the total concentration of organic compounds present and reflects the importance of having an easy-to-measure, general indicator of the approximate level of organic contamination.

It also reflects the appeal of a parameter which has a name which sounds more fundamental than it is. In many cases, the TOC is used as an on-going monitor of change or lack of change in organic content.

When completing TOC analysis, the following is measured:

- TC: Total Carbon.
- TIC: Total Inorganic Carbon.
- POC: Purgeable Organic Carbon.
- NPOC: Non-Purgeable Organic Carbon.
- DOC: Dissolved Organic Carbon.
- NDOC: Non-Dissolved Organic Carbon.

To calculate TOC, you can subtract the total amount of inorganic carbon from total carbon found. Alternatively, you can add Purgeable and Non-Purgeable Organic Carbon, or Dissolved and Non-Dissolved Organic Carbon. As sums, they look like:

TOC = TC - TIC.

TOC = POC + NPOC.

TOC = DOC + NDOC.

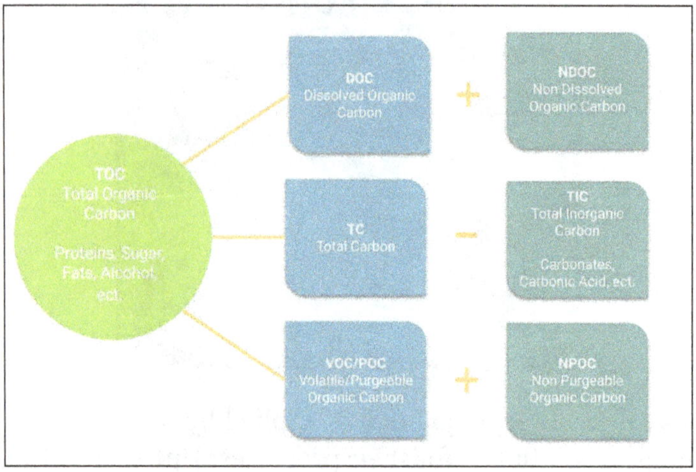

How is Total Organic Carbon Measured?

TOC is measured at very different concentrations in a very wide range of systems. Levels vary widely within each type, but, broadly, they range from sub-ppb levels in ultra-pure water for laboratory and microelectronic applications up to hundreds of ppm in effluents and process streams. For many of these systems, the TOC alone does not provide enough information. The carbon-containing compounds can be present in different forms and the proportions of each can be critical.

Dissolved organic carbon (DOC) is generally taken as that which will pass through a 0.45um filter. Large particle size TOC is classified as particulate or non-dissolved (NDOC). About 50 to 75% of DOC in natural waters is in the form of polymeric organic acids - fulvic and humic acids.

About 10% of the TOC is in colloids, mainly humic acids and various minerals. A further 10 to 20% are small molecules from the decomposition of organic matter.

Table: Total Organic Carbon levels in various types of water.

Water	Typical TOC (Range)	Typical TOC present as Particulates
Bog	33 ppm (10 to 60)	4 ppm
Marsh	17 ppm (10 to 60)	3 ppm

Eutropic lake	12 ppm	3 ppm
Oligotrophic lake	2.2 ppm	0.2 ppm
River	7.0 ppm (1 to 10)	2.5 ppm
Precipitation	1.1 ppm	0.1 ppm
Ground Water	700 ppb	
Sea Water	500 ppb	50 ppb
Waste Water	up to 1000 ppm	
Process Waters	very wide range	
Drinking Water	100 ppb to 10 ppm	
Purified Water	1 ppb to 500 ppb	
Ultrapure Water	0.1 ppb to 10 ppb	

Total Organic Carbon Analysis

TOC is universally measured by oxidizing the organic compounds present to forms which can be quantified.

A variety of oxidation and detection methods are used depending on the nature and concentration of TOC being measured and the analytical requirements (e.g. speed, sensitivity).

- High-temperature combustion at 1,200 °C in an oxygen-rich atmosphere. The CO_2 produced is passed through scrubber tubes to remove interferences and measured by non-dispersive infrared absorption (NDIR).
- High-temperature catalytic oxidation at 680 °C in an oxygen-rich environment inside tubes filled with a platinum catalyst followed by NDIR.
- Thermo-chemical oxidation with heat and a chemical oxidizer, usually a persulphate.
- Photo-chemical oxidation with UV and a chemical oxidizer, usually a persulphate.
- Photo-oxidation by ultra-violet (UV) light alone or with a catalyst. The UV oxidation method offers the most reliable, low maintenance method of determining TOC in ultra-pure waters.

The combustion methods (1 & 2) are mainly used for higher (ppm or greater) TOC concentrations or where there are high levels of particulates.

Persulphate oxidation, enhanced by UV or heat, is widely used for laboratory TOC determination in many types of water from potable to pharmaceutical and electronic grades. The CO_2 produced is usually measured by NDIR or by the change of conductivity that it produces when dissolved in a separate stream of pure water.

To exclude the effect of other oxidation products the gas may be passed through a membrane. Good oxidation can be achieved but a compensation method is needed to account for the blank from the reagent. TOC at ppb levels can be detected. All TOC analyzers which measure CO_2 will include CO_2 from bicarbonates and carbonates unless this inorganic carbon (IC) is accounted for. IC can be removed by acidifying the sample to a pH value of two or less to release IC as CO_2 which can be measured or vented to waste.

The remaining non-purgeable TOC (NPOC) contained in the liquid is then oxidized releasing CO_2 which is sent to the detector for measurement. The situation is somewhat different for high purity waters with a low conductivity. Water with a sufficiently low conductivity cannot contain significant concentrations of bicarbonates or carbonates (or other soluble salts) and no correction for IC is needed. All such trace-level measurements must be done on-line.

A number of TOC monitors have evolved to meet the need for rapid monitoring of low (ppb) TOC levels in high purity laboratory water systems. A rapid response is needed to ensure that the results are available and relevant to the relatively small volumes of water being dispensed. These monitors, generally, measure the conductivity of the water before and after oxidation; the change is calibrated against TOC content. Due to time limitations, oxidation is not always complete but is sufficient for monitoring purposes.

TOC Analysis or in the Laboratory

The great majority of samples for TOC are taken and analyzed in a laboratory. Where sub-divisions of TOC are needed, such as NDOC, samples are prepared before analysis. On-line analyses are used for higher level TOC analyses when the required frequency or speed of analysis makes it preferable. On-line measurements are also essential for the measurement of TOC levels below 50ppb to avoid contamination.

This contamination can be from extraneous TOC in the environment or containers but, more seriously, from carbon dioxide in the air which will rapidly dissolve in pure water. Carbon dioxide interferes with many of the techniques used to monitor trace TOC.

Water for TOC Analyzers and Monitors

Clearly, the purity of water needed for preparation of standards and blanks, for rinsing systems and cleaning components is very dependent on the concentrations being measured.

For many, higher level, applications, carried out in the laboratory or on-line, type II water will be more than adequate. Consistent quality will be more important than very high purity. For trace TOC determination (<50 ppb) which need to be carried out on-line, the water purity requirements are extremely stringent. As TOC will include carbon from all types of compound, they must all be virtually absent. Ultrahigh purity water with built-in TOC monitoring is strongly recommended.

CHEMICAL OXYGEN DEMAND

Chemical Oxygen Demand (COD) analysis is a measurement of the oxygen-depletion capacity of a water sample contaminated with organic waste matter. Specifically, it measures the equivalent amount of oxygen required to chemically oxidize organic compounds in water.

COD is used as a general indicator of water quality and is an integral part of all water quality management programs. Additionally, COD is often used to estimate BOD (Biochemical Oxygen Demand) as a strong correlation exists between COD and BOD, however COD is a much faster, more accurate test.

Usage of the Current COD Methods

The traditional chemical oxygen demand (COD) analysis method is the wet chemistry method. This involves a two hour digestion at high heat under acidic conditions in which potassium dichromate acts as the oxidant for any organic material present in a water sample. Silver sulfate is present as the catalyst and mercuric sulfate acts to complex out any interfering chloride. Following the digestion, the extent of oxidation is measured through indirect measurement of oxygen demand via electrons consumed in the reduction of Cr6+ to Cr3+. This can be done by titration or spectrophotometry.

A COD test can be used to easily quantify the amount of organics in water. The most common application of COD is in quantifying the amount of oxidizable pollutants found in surface water (e.g. lakes and rivers) or wastewater. COD is useful in terms of water quality by providing a metric to determine the effect an effluent will have on the receiving body, much like biochemical oxygen demand (BOD).

The basis for the COD test is that nearly all organic compounds can be fully oxidized to carbon dioxide with a strong oxidizing agent under acidic conditions. The amount of oxygen required to oxidize an organic compound to carbon dioxide, ammonia, and water is given by:

$$C_nH_aO_bN_c + \left(n + \frac{a}{4} - \frac{b}{2} - \frac{3}{4}c\right)O_2 \rightarrow nCO_2 + \left(\frac{a}{2} - \frac{3}{2}c\right)H_2O + cNH_3$$

This expression does not include the oxygen demand caused by nitrification, the oxidation of ammonia into nitrate:

$$NH_3 + 2O_2 \rightarrow NO_3^- + H_3O^+$$

Dichromate, the oxidizing agent for COD determination, does not oxidize ammonia into nitrate, so nitrification is not included in the standard COD test.

The International Organization for Standardization describes a standard method for measuring chemical oxygen demand in ISO 6060 .

Using Potassium Dichromate

Potassium dichromate is a strong oxidizing agent under acidic conditions. Acidity is usually achieved by the addition of sulfuric acid. The reaction of potassium dichromate with organic compounds is given by:

$$C_nH_aO_bN_c + dCr2O_7^{2-} + (8d + c)H^+ \rightarrow nCO_2 + \frac{a+8d-3c}{2}H_2O + cNH_4^+ + +2dCr^{3+}$$

where $d = 2n/3 + a/6 - b/3 - c/2$. Most commonly, a 0.25 N solution of potassium dichromate is used for COD determination, although for samples with COD below 50 mg/L, a lower concentration of potassium dichromate is preferred.

In the process of oxidizing the organic substances found in the water sample, potassium dichromate is reduced (since in all redox reactions, one reagent is oxidized and the other is reduced), forming Cr^{3+}. The amount of Cr^{3+} is determined after oxidization is complete, and is used as an indirect measure of the organic contents of the water sample.

Measurement of Excess

For all organic matter to be completely oxidized, an excess amount of potassium dichromate (or any oxidizing agent) must be present. Once oxidation is complete, the amount of excess potassium dichromate must be measured to ensure that the amount of Cr^{3+} can be determined with accuracy. To do so, the excess potassium dichromate is titrated with ferrous ammonium sulfate (FAS) until all of the excess oxidizing agent has been reduced to Cr^{3+}. Typically, the oxidation-reduction indicator ferroin is added during this titration step as well. Once all the excess dichromate has been reduced, the ferroin indicator changes from blue-green to a reddish brown. The amount of ferrous ammonium sulfate added is equivalent to the amount of excess potassium dichromate added to the original sample. Ferroin indicator is bright red from commercially prepared sources, but when added to a digested sample containing potassium dichromate it exhibits a green hue. During the titration the color of the indicator changes from a green hue to a bright blue hue to a reddish brown upon reaching the endpoint. Ferroin indicator changes from red to pale blue when oxidized.

Preparation of Ferroin Indicator Reagent

A solution of 1.485 g 1,10-phenanthroline monohydrate is added to a solution of 695 mg $FeSO_4 \cdot 7H_2O$ in distilled water, and the resulting red solution is diluted to 100 mL.

Calculations

The following formula is used to calculate COD:

$$COD = \frac{8000(b-s)n}{\text{sample volume}}$$

where *b* is the volume of FAS used in the blank sample, *s* is the volume of FAS in the original sample, and *n* is the normality of FAS. If milliliters are used consistently for volume measurements, the result of the COD calculation is given in mg/L.

The COD can also be estimated from the concentration of oxidizable compound in the sample, based on its stoichiometric reaction with oxygen to yield CO_2 (assume all C goes to CO_2), H_2O (assume all H goes to H_2O), and NH_3 (assume all N goes to NH_3), using the following formula:

$$COD = (C/FW)\cdot(RMO)\cdot 32$$

Where ,

C = Concentration of oxidizable compound in the sample,

FW = Formula weight of the oxidizable compound in the sample,

RMO = Ratio of the # of moles of oxygen to # of moles of oxidizable compound in their reaction to CO_2, water, and ammonia

For example, if a sample has 500 Wppm (Weight Parts per Million) of phenol:

$$C_6H_5OH + 7O_2 \rightarrow 6CO_2 + 3H_2O.$$

$$COD = (500/94)\cdot 7\cdot 32 = 1191 \text{ Wppm.}$$

Inorganic Interference

Some samples of water contain high levels of oxidizable inorganic materials which may interfere with the determination of COD. Because of its high concentration in most wastewater, chloride is often the most serious source of interference. Its reaction with potassium dichromate follows the equation:

$$6Cl^- + Cr_2O_7^{2-} + 14H^+ \rightarrow 3Cl_2 + 2Cr^{3+} + 7H_2O$$

Prior to the addition of other reagents, mercuric sulfate can be added to the sample to eliminate chloride interference.

The following table lists a number of other inorganic substances that may cause interference. The table also lists chemicals that may be used to eliminate such interference, and the compounds formed when the inorganic molecule is eliminated.

Inorganic molecule	Eliminated by	Elimination forms
Chloride	Mercuric sulfate	Mercuric chloride complex
Nitrite	Sulfamic acid	N_2 gas
Ferrous iron	-	-
Sulfides	-	-

BIOCHEMICAL OXYGEN DEMAND

Biochemical oxygen demand (BOD) is the most important and widely used environmental index for monitoring organic pollutants in wastewater. The current international standard method is the five-day BOD assay (BOD_5). The test measures the amount of dissolved oxygen required for the microbial oxidation of carbonaceous organic material under specified conditions. The test has its widest application in measuring waste loadings to treatment plants and in evaluating the BOD-removal efficiency of such treatment systems.

Practical difficulties exist associated with the BOD_5 test, including sensitivity to temperature and oxygen concentration, impact of toxins, lack of stoichiometric validation, and the need to dilute samples. The test is labor intensive with variable results, but the major drawback continues to be the five-day measurement time.

For all its shortcomings and the availability of a number of related measurement techniques, the BOD_5 assay remains the preferred test for reporting the oxygen requirements of wastewaters, effluents, and polluted waters, in part, because of its international regulatory status, but also because it uniquely simulates the microbial decomposition of organic wastes in the environment, thereby approximating the response of natural ecosystems.

References

- Trends-in-the-analysis-of-persistent-organic-pollutants-POPs-in-human-blood-234036316: researchgate.net, Retrieved 23 February, 2020
- Analysis-persistent-organic-pollutants-using-pressurized-liquid-extraction-and-automated-column-chro: chromatographyonline.com, Retrieved 11 May, 2020
- Remote-zones-air-quality-persistent-organic-pollutants-sources-sampling-and-analysis, air-quality-monitoring-assessment-and-management: intechopen.com, Retrieved 08 August, 2020
- Comprehensive-analysis-persistent-organic-pollutants-complex-matrices-48-using-gc-high-performance-t: spectroscopyonline.com, Retrieved 07 April, 2020

Effects of Organic Pollutants

CHAPTER 5

Organic pollutants have adverse health effects on humans which include cancer, immune system suppression, decrements in cognitive and neurobehavioral function, disruption of sex steroid and thyroid function, and various chronic diseases, such as hypertension, cardiovascular disease, and diabetes. This chapter discusses the effects of organic pollutants in detail.

ENVIRONMENTAL IMPACT OF PERSISTENT ORGANIC POLLUTANTS

There is a profound relationship among air, water, animals and humans in the environment. The effects on these identities directly or indirectly disturb the environment. Therefore, the environment is being affected by POPs through biotic, abiotic, social, cultural and technological interferences. The pollution of POPs degrade ecological balance; threatening the environment and health of all organisms. POPs accumulated in marine animal leading to their death. This results into imbalance in sea ecology.

Sea, Great Lakes and the Arctic Sea in 1980s and 1990s. These areas suffered from disturbance in aquatic ecosystems, which were having POPs concentrations in different species. Besides, central and Eastern Europe and Russia also had similar environmental disorders. Briefly, entire ecosystem is being affected by POPs. The insects, birds, mammals and other organisms are poisoned either directly or indirectly. The environmental effects include shift in sex ratios, wildlife cancer, impaired fertility and other physical abnormalities. Industrial revolution of the 19th and 20th centuries increased POPs greatly. Some POPs are volatile and get accumulated on dust and aerosol. These reach to our habitats and get inhaled into our body. Walker described inhalation and ingestion of dust and air as major sources of POPs.

Adeola described the environmental effects of POPs. Wong et al. reviewed the impacts of POPs on the environment of China and North America. Letcher et al. reviewed the effect of POPs on several top trophic level species i.e. seabirds (e.g., glaucous gull (Larus hyperboreus), polar bears (Ursus maritimus), polar fox (Vulpes lagopus), Arctic charr (Salvelinus alpinus) and sled dogs (Canis familiaris). The effects were discussed in terms of biochemical processes (immune and endocrine system, pathological changes in tissues and reproduction and development) and tissue residue levels of POPs (PCBs,

DDTs, CHLs, PBDEs, PFCAs and PFSAs). The authors discussed the effects on Arctic organisms, polar bears and glaucous gulls. Besides, POPs effect on the environment, ecology and physiology stressors were also highlighted. For example, changes in food style, fattening and emaciation were also observed. The other factors such as climate changes (temperature, precipitation, food web, nutrition changes), disease and species invasion were described. The authors recommended future research and better understanding of POPs effects on animals. Noyes et al. reviewed the effects of climate change due to POPs.

The changes on the environmental parameters such as temperature, precipitation and salinity were discussed. The climatic changes were observed in food webs, lipid dynamics, ice and snow melt and organic carbon cycling. Some other reviews also discuss the environmental effects of POPs. Mostly POPs exist in the environment for more than 20 years with few as long as century. DDT, endrin, HBC, mirex, polychlorinateddibenzo-p-dioxins and furans remain in the environment for 10–20 years. POPs show a process called grasshopper effect involving cyclic volatilizations and condensations. These migrate from warmer climate to cooler one. Sometimes, these settle down under extremely cool conditions but evaporate when the temperature rises. Sometimes, the presence of this cyclic migration of POPs creates global warming problem. The authors described the effects of polychlorinated biphenyls, dichloro-phenyl-dichloroethylene, hexachlorobenzene and oxychlordane on hormonal disturbance in Arctic mammals and seabirds; leading to ecological disturbance. The effects were observed in thyroid hormone, sex steroid hormones and cortisol; resulting into behavioral and morphological changes. Kong et al. discussed the effects of POS on the environment by modeling approach using the POPCYCLING-Baltic multimedia chemical fate model. The authors predicted changes in temperature, wind speed, precipitation and carbon aerosols of POPs in the Baltic Sea. Komprda et al. used model assessment for studying influence of climate and land use changes due to emission of organochlorine pesticides. The effects of plowing influence of land cover, temperature of soil, and seasonal changes of air layer stability on emission were studied. The results indicated that emission occurred due to style of agricultural activities and temperature variations.

MALE REPRODUCTIVE HEALTH

The possible effects of exposure to environmental contaminants on male reproductive function have been addressed in a number of epidemiological studies. Although these studies suggest some effects on male reproductive health, the results across studies are often conflicting.

There are considerable numbers of animal studies that indicate adverse effects of environmental toxicants on male reproductive health in both experimental and wildlife settings but it should be noted that the exposure levels causing reproductive toxicity in most of these studies were several times higher than the human exposure levels measured in the general population.

Special attention has been given to the potential adverse effects on human reproductive function of compounds that are suspected to interfere with the endogenous regulation of hormonal action and thus are suspected to interfere with the development and function of reproductive hormone- regulated processes such as genital development, puberty onset and sperm production. Due to interference with hormone receptors or steroid-producing or degrading enzymes, even low concentrations of these compounds may disturb endocrine-regulated processes.

Several of these compounds are suspected to affect male reproductive health. The major groups include phthalates and bisphenol A, which have typically been used in the plastic industry, polychlorinated biphenyls (PCBs), which were used extensively in transformers, insulating condensers, as plasticizers, insulators, paints and flame retardants and organochlorine pesticides such as dichlorodiphenyltrichloroethane (DDT) and hexachlorobenzene (HCB).

In recent years, other halogenated organic compounds have received increased attention, including brominated flame retardants (BFRs) such as polybrominated diphenyl ethers (PBDEs) and the perfluorinated compounds (PFCs) that have been used as surfactants in industrial processes and extensively as oil and water repellents for consumer products. In addition to the organic compounds, certain metals (e.g. mercury, lead and cadmium) have been suspected to interfere with male reproductive health, as effects on male reproductive function have been demonstrated at high exposure levels in occupational and experimental settings.

The biopersistent compounds, including the halogenated organic compounds, have been a subject of particular concern, as they often have half-lives in the human body of more than 5 years and are bioaccumulative. Several of these compounds can be detected in human serum around the globe, and a large number of these persistent organic pollutants (POPs) have been regulated to avoid further bioaccumulation in humans.

Several methods have been developed to study male fertility. Semen quality is often used as a marker of male fertility, and the probability of achieving pregnancy (fecundability) has been demonstrated to be highly dependent on concentrations of sperm of up to approximately 40 million ml^{-1}. In addition to sperm concentration, sperm morphology and motility are part of the standard evaluation of semen quality and have also been shown to be associated with fecundability independently of sperm concentration. Other attempts have been made to develop alternative measures of semen quality, including the development of different assays for the assessment of sperm chromatin integrity. Such measures have also been demonstrated to be associated with fecundability independently of standard sperm parameters.

Malformations of the male genital organs (cryptorchidism and hypospadias) have been studied as markers of male reproductive disturbances, as these outcomes can usually be detected at birth. Although they can be surgically corrected, these malformations are related to reduce male fertility in adulthood, especially if left untreated. In addition, male anogenital distance has recently been demonstrated to be related to semen quality and chance of fatherhood and may be used as a marker of male fecundability. Furthermore, the endocrine-regulated timing of puberty has been suggested to be affected by environmental contaminants, and in addition to psychosocial consequences, early puberty onset in boys has been related to the subsequent risk of testicular cancer. Finally, reproductive hormone levels are considered markers for male reproductive function.

Although the majority of previous studies have addressed adult exposure and reproductive health, theoretical considerations supported by animal studies and a number of recent epidemiological studies suggest that exposure during the fetal period may have more severe long-term consequences for male reproductive health than exposure in adulthood, where the observed effects are more likely to be reversible.

Persistent Organic Pollutants and Male Reproductive Health

Human exposure characterization generally, human exposure to POPs has declined because the production of most POPs has ceased or been significantly reduced. However, some of the more recently developed POPs are still in use or have been phased out only recently.

The majority of nonoccupationally exposed general populations are exposed to POPs through dietary intake of environmentally contaminated food, but exposure may also result from drinking water, air or dermal contact with POPs. Additionally, children are exposed to maternal POP levels during intrauterine development and postnatally through breast feeding and inhalation of dust while crawling. Moreover, for PFCs, which unlike the majority of other POPs accumulate in the blood and liver instead of adipose tissue, indirect exposure as a result of biotransformation/degradation of residual or commercial fluorochemical starting materials is also thought to be a significant source of exposure.

POPs enter the environment from a number of industrial sources, including industrial waste water, incinerating plants, power stations and landfills and they are not removed by conventional waste water treatment processes. POPs can be found in water, soil and sediment, and are transported across the globe through the atmosphere and by fresh and marine water currents far from their original production or usage sites.

PCB-138, -153 and -180 are the most common PCB congeners in biological samples, and 2,2',4,4',5,5'-hexachlorobiphenyl(CB-153) is often used as a proxy for total PCB exposure because $\sum$PCB and CB-153 are highly correlated. Many studies have suggested that fish consumption is a strong predictor for higher PCB concentrations, but the intake of dairy products and meat has also been associated with higher levels of PCBs.

Studies on Scandinavian background exposure levels of CB-153 performed in the 2000s demonstrated this level to be in the range of 50–68 ng g^{-1} lipid which is similar to mean levels in the USA during the same time period (44 ng g^{-1} lipid). Temporal trends in human sera suggest that the levels of perfluorinated compounds such as perfluorooctanoic acid (PFOA) and perfluorooctanesulfonic acid (PFOS) peaked around the year 2000 and are currently declining. Generally, PFC exposure is high in industrialized countries, but it is also high in the Arctic due to bioaccumulation in the food chain. Recent mean serum levels of PFOA and PFOS, as measured in Red Cross blood donors in 2010 in the USA, were 2.4 and 8.3 ng ml^{-1}, respectively, and between 2001 and 2010, these levels declined 48% for PFOA and 76% for PFOS, respectively. Such levels are comparable to background exposure levels that were measured in Danish men in 2008–2009 (mean PFOA: 3.5 ng ml^{-1} and mean PFOS: 8.5 ng ml^{-1}).

In 2003 to 2005, mean DDT and dichlorodiphenyldichloroethylene (p, p'-DDE) serum levels of South African men living in DDT-sprayed areas were 90.2 and 215.5 µg g^{-1} lipid, respectively; whereas, serum levels of p, p'-DDE in general populations exposed to p, p'-DDE as a result of the agricultural use of DDT were 1270 ng g^{-1} lipid (Ukraine) and 580 ng g^{-1} lipid (Poland) in the period 2002–2004 and 275 ng g^{-1} lipid in male partners of infertile couples in the period 2000–2001 (USA). The mean serum levels of men from southern and northern Norway in 2001 were somewhat lower: 81 and 66 ng g^{-1} lipid, respectively.

Although serum levels of many of the well-described POPs are now declining, the concentrations of some of the newer compounds, such as perfluorohexanesulfonic acid (PFHxS), perfluorobutanesulfonic acid (PFBS), perfluorononanoic acid (PFNA), perfluorodecanoic acid (PFDA) and perfluoroundecanoic acid (PFUnA) or their precursors, are increasing. Thus, POPs found in the environment today may still be at levels that are potentially harmful to male reproductive health.

Mechanisms

Experimental data suggest that POPs interact with steroid homeostasis and affect hormonal balance through mechanisms involving steroid receptor binding and/or the disruption of steroid biosynthesis or metabolism. Hence, many POPs are regarded as potential endocrine disrupting compounds. Depending on the congener type, PCBs have been demonstrated to exhibit dioxin-like effects, with effects exerted through the aryl hydrocarbon receptor (AhR), in addition to estrogenic, antiestrogenic and antiandrogenic effects. The major DDT degradation product p, p'-DDE is regarded as an androgen receptor antagonist, whereas animal studies suggest both estrogenic and antiandrogenic effects of BFRs.

Adult Exposure and Reproductive Health

Due to the potential endocrine-disrupting effects of POPs, an extensive number of epidemiological studies have attempted to clarify whether POPs pose a risk to human male reproductive health.

Polychlorinated Biphenyls

The potential harmful effects of PCBs on male reproductive function have been studied extensively. Overall, studies of exposure to PCBs during adulthood indicate some association between PCB and lower sperm motility and to some extent, decreased sperm DNA chromatin integrity and lower levels of free testosterone.

p, p '-Dichlorodiphenyldichloroethylene

In a South African study of a healthy cohort of 311 men living in an endemic malaria area, the mean p, p'-DDE serum level of men living in DDT-sprayed houses was 239 $\mu g\ g^{-1}$ lipid compared with 99.5 $\mu g\ g^{-1}$ lipid for men whose houses were not sprayed. The study suggested an inverse association between p, p'-DDE exposure and semen volume, total sperm count and computer-assisted sperm analysis mean motility, which was partly corroborated by another study of 116 men environmentally exposed to high DDT/p, p'-DDE levels in Chiapas, Mexico. Here, crude results suggested that increasing plasma p, p'-DDE concentrations were inversely associated with sperm motility and were positively associated with sperm tail defects and inadequate sperm chromatin condensation. Similar to the results on PCBs, background levels of p, p'-DDE have also been found to be associated with reduced sperm motility. This finding was corroborated by crude but not adjusted analyses by Rignell-Hydbom et al. However, other studies did not find any statistically significant associations between p, p'-DDE and sperm motility.

Contradictory to the hypothesized adverse effect of p, p'-DDE exposure on sperm quality, a few studies have identified positive associations between p, p'-DDE exposure and sperm concentration. A weak association between p, p'-DDE and sperm concentration was found in the northern population of a Norwegian study, which also reported a positive association between PCB exposure and sperm concentration. In the study by Toft et al., a positive association between p, p'-DDE exposure and sperm concentration was observed in a Swedish cohort. This association, however, was suggested to be a chance finding caused by lower than normal sperm concentration in the group with low exposure. Several studies have suggested that p, p'-DDE is not related to sperm morphology, sperm DNA integrity, or reproductive hormone levels, although the study by Giwercman et al., suggested some effects of p, p'-DDE exposure on adult male reproductive hormone levels.

Perfluorinated Compounds

Results from epidemiological cross-sectional studies of the potential associations between PFCs in adulthood and male reproductive health are conflicting. Some studies suggest effects on sperm morphology, motility and reproductive hormone levels, whereas others report no observed effects. Two cross-sectional studies of men occupationally exposed to PFOA in 1993 and 1995 in the USA found no associations between high serum levels of PFOA and altered reproductive hormone levels.60 In contrast, studies on background exposure levels of PFOA and PFOS suggest effects on reproductive hormones levels.

Thus far, only one study has investigated possible associations between PFC exposure and sperm DNA integrity. This study of a cohort of fertile men found that serum PFOA but not PFOS, PFHxS or PFNA in men from Greenland was associated with an increased percentage of (terminal deoxynucleitidyl transferase dUTP nick-end labeling (TUNEL))-positive spermatozoa (increased level of DNA strand breaks), but not an elevated DNA fragmentation index, as assessed by the sperm chromatin structure assay. Additionally, the study suggested that higher levels of spermatozoa positive for the proapoptotic marker Fas were associated with a higher exposure to PFOS in Polish men, whereas only tendencies towards elevated % Fas positivity were indicated for the other regions investigated. Overall, the study did not suggest strong associations between PFOS, PFOA, PFHxS or PFNA and sperm DNA damage or apoptosis. Based on the few studies that have evaluated the possible reproductive health effects of adult exposure to PFCs, it appears that the direction of associations between PFCs and reproductive hormones are not consistent, but that sperm morphology may be targeted by PFC exposure.

Brominated Flame Retardants, Dioxin A and Hexachlorobenzene

Knowledge is sparse as to how adult exposure to BFRs, dioxins and HCBs affects male reproductive health. However, a Japanese study of 10 young men indicated inverse associations between BDE-153 and sperm concentration and testis size, and a Canadian study of 52 men recruited at an infertility clinic indicated adverse effects of BDE-47, BDE-100 and the sum of measured PBDE congeners (BDE-47, BDE-99, BDE-100 and BDE-153) on sperm motility, but no relation with other semen parameters. Finally, a study on 62 men from the USA indicated positive associations between house dust levels of penta-BDEs and serum levels of free and total testosterone, estradiol and sex hormone binding globulin (SHBG) and inverse associations with follicle stimulating hormone (FSH). In the same study, house dust octa-BDEs were positively associated with luteinizing hormone (LH) and testosterone, and finally, deca-BDEs were inversely associated with testosterone. This study expanded a previous report from the same group indicating inverse associations between measured PBDE congeners (47, 99 and 100) in dust samples and free testosterone, LH and FSH and positive associations with SHBG and inhibin B.

Exposure to dioxins during adulthood does not appear to cause long-term effects on semen quality or reproductive hormone levels when measured 20 years later, as suggested in a study of Italian men exposed to high 2,3,7,8-tetrachlorodibenzo-p-dioxin levels when they were 18–26 years of age as a result of the Seveso explosion in 1979. In addition, HCB measurements in adulthood have not been related to changes in sperm quality, sperm DNA integrity or reproductive hormone levels.

Concluding Remarks on Adult Exposure and Persistent Organic Pollutants

Based on the effects described in these studies, it appears that exposure to PCBs during adult life primarily affects sperm motility, whereas the perfluorinated compounds included here appear to primarily target sperm morphology.

Fetal Exposure and Reproductive Malformations at Birth

The possible effects on cryptorchidism and hypospadias of in utero exposure to DDT and p, p'-DDE have been studied in two nested case-referent studies in the USA: the Child Health and Development Study, including 75 cases of cryptorchidism, 66 cases of hypospadias and 283 controls; and the Collaborative Perinatal Project (CPP), including 219 cases of cryptorchidism, 199 cases of hypospadias and 552 controls.67 Both studies included pregnancies in the period from 1959 to 1966, and both found slightly but not significantly increased risk of cryptorchidism and hypospadias, with odds ratios of approximately 1.3 at the highest maternal p, p'-DDE exposure level compared to the lowest maternal p, p'-DDE exposure level. Furthermore, Swedish study on 237 hypospadias cases compared with a similar number of controls suggested that increased p, p'-DDE and HCB exposure, but not PCB exposure may be associated with an increased risk of hypospadias.

No association between placenta concentrations of PCB and dioxins and the risk of cryptorchidism was observed in a Danish-Finnish study. Although no evidence of an association between PCB or HCB exposure and cryptorchidism was suggested, an increased risk of hypospadias was suggested at high PCB exposure levels in the CPP in the USA.

A few studies have used breast milk as a proxy for in utero exposure, and a French study found an association between breast milk PCB concentrations and cryptorchidism. Additionally, a Danish-Finnish study found an increased risk of cryptorchidism at higher breast milk levels of PBDE, but did not find any associations with placental PBDE levels. Based on such small and frequently nonstatistically significant effects, it remains inconclusive whether persistent organic compounds at environmental exposure levels increase the risk of hypospadias and cryptorchidism, but it seems that, at most, a small increased risk can be ascribed to exposure to these compounds.

Another outcome indicating feminization of males is decreased anogenital distance, as females have a shorter anogenital distance than males. This outcome was more sensitive to exposure to environmental antiandrogens than hypospadias and cryptorchidism in animal studies. Thus, although the clinical consequences of reduced anogenital distance are probably limited, it may be a sensitive marker of endocrine disturbances in humans. In the USA a smaller study including 37 male offspring indicated reduced anogenital distance at higher p, p'-DDE exposure. However, a larger study among 781 mother-child pairs in Chiapas, Mexico, of which 29% reported living in DDT-sprayed homes, indicated no association between p, p'-DDE exposure and anogenital distance or penile length, suggesting that even high exposure to p, p'-DDE does not seem to have a significant impact on these outcomes in humans.

Perinatal Exposure and Reproductive Health in Adulthood

A large number of studies have investigated possible male reproductive health effects of exposure to POPs during adult life. However, only a few studies have had the opportunity to investigate the very core of the endocrine disruptor hypothesis, namely, the possible long-term effects of exposure to compounds with endocrine-disrupting effects during fetal life on male reproductive health.

Exposure to endocrine-modulating agents during fetal life is of particular concern because an imbalance of the fetal hormonal environment may affect the normal development of male reproductive organs, which may cause long-term effects on the male reproductive system.

The Yu-Cheng oil-disease in Taiwan region in 1979, where over 2000 men and women were accidently contaminated with high levels of PCBs and their pyrolytic products (e.g. polychlorinated dibenzofurans), has provided valuable knowledge regarding how high exposures to POPs during fetal development affect adult human male reproductive health. Comparisons between prenatally exposed men (n = 12) and unexposed controls (n = 23) demonstrated increased abnormal morphology and a reduced percentage of progressively motile and rapidly motile spermatozoa in prenatally exposed men compared with controls. Additionally, spermatozoa from exposed men exhibited reduced oocyte penetration capacity compared with controls.

Another study that assessed the effects of accidental perinatal exposure to dioxin as a result of the trichlorophenol plant explosion near Seveso, Italy indicated permanent effects on sperm quality. The extrapolated median serum 2,3,7,8-tetrachlorodibenzo-p-dioxin concentration of the mothers at the time of conception was 26 parts per trillion. In this study, the background levels of dioxin exposure were assumed to be 10 parts per trillion, and serum concentrations in breastfed sons were considered to be twofold that of the mothers. The authors found that the 39 men exposed in utero and postnatally through breastfeeding exhibited significantly lower sperm concentrations, total sperm counts, numbers of total motile spermatozoa and inhibin B concentrations and higher FSH concentrations compared with the controls. There were no differences in any of the semen parameters or reproductive hormones between the formula-fed men only exposed to dioxin in utero (n = 18) and the controls. Hence, the male reproductive system seems to be sensitive to perinatal dioxin exposure rather than in utero exposure alone.

A recent follow-up study of a cohort of 169 Danish men who participated in a study in 2008-2009 when they were 19 to 21 years-old addressed the potential long-term effects of in utero exposure to background levels of the perfluorinated compounds PFOA and PFOS on adult male reproductive health. The study indicated inverse associations between PFOA and sperm concentrations and total sperm counts and that higher exposure to PFOA during fetal development was associated with higher levels of LH and FSH in adulthood, whereas no associations between PFOS and any of the semen parameters or reproductive hormones were found.

RISK OF METABOLIC DISEASES

The incidence of obesity and diabetes has dramatically increased worldwide. From 1980 to 2008, the global prevalence of obesity has doubled in both men and women, whereas the number of people with diabetes increased from 153 million in 1980 to 347 million in 2008. Report realised in the European Union (EU) highlights that overweight and obesity affect more than 50% of the adult population whereas diabetes is now affecting over 30 million people. In the United States (US), the prevalence of obesity among children and adolescents has almost tripled since 1980, and 12% of children aged 2 through 5 years were obese in 2009-2010. In addition to enhancing the risk of premature death, diabetes and obesity are major causes of multiple complications, including hypertension, cardiovascular diseases, asthma, blindness, limb amputation, and sleep apnea, that generate enormous economic costs for both health care and loss of productivity to society. Total annual economic cost of diabetes in the US reached $132 billion in 2002, representing around 11% of the US health care expenditure, and increased to $174 billion in 2007. In Europe, obesity-related healthcare was estimated to reach up to 10.4 billion Euros.

Although they are the focus of intense investigations, the origins of metabolic diseases have remained poorly understood. While there is certainly genetic predisposition, the rapid and explosive incidence of obesity and type 2 diabetes cannot be due to genetic modifications in the population. Indeed, the propagation of these diseases is simply too fast to lie at the feet of genetic changes. Rather, environmental factors like physical inactivity and consumption of high-energy diets are likely to be important contributors. However, the view that these two conventional risk factors alone are the primary variables explaining the epidemic of metabolic diseases is being challenged as far too simplistic.

Persistent organic pollutants (POPs), including dioxins, furans, polychlorinated biphenyls (PCBs), and organochlorine pesticides, are chemicals mainly created by industrial activities, either intentionally or as by-products. Because of their ability to resist environmental degradation, these substances are omnipresent in food products, and found all around the world, even in areas where they have never been used like

Antarctica. Thus, virtually all humans are daily exposed to POPs. In the general population, exposure to POPs comes primarily from the consumption of animal fat like fatty fish, meat and milk products; the highest POP concentrations being commonly found in fatty fish.

POPs are lipophilic chemicals that can pass through biological phospholipid membranes and bio-accumulate in fatty rich tissues of humans. Initially described for their deleterious effects on reproductive function and carcinogenicity, there is now growing body of evidence showing that exposure to POPs leads to metabolic diseases. First, many epidemiological studies performed in the US, Europe and Asia have led to the common finding that there is an increased body burden of POPs in people with diabetes. For instance, diabetes has been associated with elevated serum levels of 2,3,7,8-tetrachlorodibenzodioxin (TCDD), 2,2',4,4',5,5'-hexachlorobiphenyl (CB153), coplanar polychlorinated biphenyls (PCBs; CB77, CB81, CB126, CB169), p,p'-diphenyldichloroethene (DDE), oxychlordane and trans-nonachlor. In addition, bio-accumulation of PCBs has been linked to non-alcoholic fatty liver disease (NAFLD) and elevated blood pressure. Recently, prospective studies have furthermore shown an increased risk of diabetes in persons exposed to POPs, especially organochlorine pesticides. Second, there is evidence for a causal relationship between POP exposure and metabolic disorders linked to insulin resistance. Animals exposed to environmental levels of POP mixtures through the intake of non-decontaminated fish oil (obtained from farmed Atlantic salmon) exhibited insulin resistance, glucose intolerance, abdominal obesity and NAFLD. In rats fed decontaminated crude salmon oil, which contained very low levels of POPs, these metabolic disturbances were almost absent. Importantly, the body burden of POPs in rats exposed to non-decontaminated salmon oil did not exceed those observed in humans 40-50 years of age, thereby underlying that the animals were not over-exposed to POPs. Further, the presence of POPs in farmed Atlantic salmon fillet was found to accelerate the development of visceral obesity and insulin resistance in mice. Other studies exposing rodent to single POP compound reported metabolic dysfunctions like NAFLD or increased body weight gain. Third, in vitro investigations showed that POPs are powerful disruptors of metabolic homeostasis, which is consistent with the obesogen properties of some xenobiotics. Taken together, these findings observed in humans, animals, and cell models, demonstrate that POPs are potent environmental contributors to metabolic diseases.

Current Regulation and Risk Assessment of Pops

To protect humans, the United Nations Environment Programme (UNEP), in 1997, issued an international action to reduce the global release of 12 POPs, the "dirty dozen", which included nine organochlorinated pesticides; aldrin, chlordane, dichlorodiphenyltrichloroethane (DDT), dieldrin, endrin, heptachlor, hexachlorobenzene, mirex, toxaphene, and three industrial chemicals; dioxins, furans, and PCBs. In 2001, the Stockholm Convention on POPs was ratified by 151 signatories, and the Convention entered into force three years later.

Aldrin, chlordane, DDT, dieldrin, endrin, heptachlor, mirex, and toxaphene were extensively used in agriculture as insecticides, whereas hexachlorobenzene was used as a fungicide. By 1990, these POPs were prohibited for use on crops and termite control by most European and North American countries because of their potential threat to human health. DDT has also been utilized in indoor spraying for control of vectors of malaria, dengue, visceral leishmaniasis, and Chagas disease, and its use for the purpose of controlling disease vectors has been permitted by the Stockholm Convention, in accordance with recommendations and guidelines of the World Health Organization (WHO). In its Position Statement, WHO stipulates that a sudden ban on DDT use could have significant consequences for the burden of malaria because of the absence of equally effective and efficient alternatives. The use of DDT in malaria prevention has led to intense debates because it can save many lives but DDT may, at the same time, induce adverse human health effects. In 2005, the global production of DDT used for disease vector control was estimated to reach 5,000 metric tons. Up to 160 metric tons of DDT were still produced for agriculture purposes, and the production of other organochlorine pesticides, like aldrin and dieldrin, is suspected to occur in different countries. Not surprisingly, US monitoring study revealed that DDT and its metabolites as well as endosulfan and aldrin, are still largely present in food, and daily consumed by humans.

The term "dioxins" is commonly used to refer a group of 75 polychlorinated dibenzo-p-dioxins (PCDDs) and 135 polychlorinated dibenzofurans (PCDFs) congeners among which less than 20 are believed to be toxic. Dioxin like-PCBs (dl-PCBs) refer to 12 non-ortho or mono-ortho PCBs exhibiting similar biological pattern to dioxins. In 1997, an expert committee, set up by the WHO, developed the concept of toxic equivalent factors (TEF) to facilitate risk assessment and regulatory control of exposure to mixtures of PCDDs, PCDFs, and dl-PCBs. TEF is based on in vitro and in vivo data showing that PCDD, PCDF and dl-PCB congeners induce similar toxic responses to those caused by TCDD, which is considered as the most potent toxic congener. Thus, TEF is an order of magnitude estimate of the toxicity of a compound relative to TCDD. The criterion to apply a TEF to a POP congener is the following:

- The chemical must show a structural relationship to PCDDs and PCDFs.
- The chemical must bind the aryl hydrocarbon receptor (AhR) and elicit AhR-mediated biochemical and toxic responses.
- The chemical must be ubiquitous in the food chain.

To further estimate the toxicity of POP mixtures to which humans and organisms can be exposed, TEF values are used to calculate toxic equivalent (TEQ) concentrations as follows: $TEQ = \Sigma_{n1} (PCDDs_i \times TEF_i) + \Sigma_{n2} (PCDFs_i \times TEF_i) + \Sigma_{n}3 (dl\text{-}PCBs_i \times TEF_i)$. This approach assumes that the combined effects of different PCDD, PCDF and dl-PCB congeners produce additive toxic effects.

The Stockholm Convention has led to substantial reduction of POP production. Dietary surveys have reported that the concentrations of organochlorine pesticides, dioxins, and PCBs have declined in most food products. Furthermore, many studies characterizing the temporal trends of these classical POPs in human blood showed that the concentrations of these pollutants have decreased during the last years, although the human body burden of so-called "emerging" POPs, like polybrominated diphenyl ether (PBDE) and some perfluorinated compounds, have remained stable or even increased. The present situation may therefore indicate that the current background exposure to POPs is still sufficient to perturb hormonal and metabolic homeostasis of the human body.

In general, safety guidelines have been performed by identifying the most sensitive effects of single POP, and has mainly focused on TCDD and its impacts on immunotoxicity, carcinogenotoxicity, reproductive and developmental toxicity in animals. In 2000, the Scientific Committee on Food (SCF) of the European Commission reviewed extensive data and experimental results and recommended a tolerable weekly intake (TWI) of 7 pg WHO-TEQ/kg body weight to protect consumers from the harmful effects associated with environmental contaminant exposure through consumption of food products. Only six months later, the SCF re-assessed the TWI by focusing on non-carcinogenetic endpoints only, and increased the TWI from 7 to 14 pg WHO-TEQ/kg body weight based on rat studies investigating the reproductive effects of male offspring. Below this level, contaminants are expected to have no effects or its effects are reversed by body's defence mechanisms. Today, various authorities have defined their own maximum daily exposure limits, tolerable daily intake (TDI), within the range of 1 to 4 pg WHO-TEQ/kg body weight/day. Still, although some population has TDI below national ranges, many studies have documented that exposure to dioxins and dl-PCBs to humans, especially children, can be above the recommended TDI levels.

Main limitations of current risk assessment on POPs in relation with metabolic disorders:

- TEQ and the concept of additive toxic effects: Although TEQ concept is based on additive effects of pollutants, i.e. high doses of a chemical are expected to cause greater deleterious effects than low doses, not all studies on environmental pollutants and xenobiotics support this concept. Accumulating data continue to underline the existence of a non-linear or inverted relationship between environmental pollutants and their detrimental effects; strong biological effects in low dose of exposure but weak or no effects in higher dose. For instance, exposure to low concentrations of TCDD and 1,2,3,4,7,8-hexachlorodibenzo-p-dioxin (HxCDD) increased body weight gain in rats whereas at higher doses of exposure, both TCDD and HxCDD decreased body weight. Consistently, low doses of CB-77 and TCDD were found to promote adipocyte differentiation whereas higher doses resulted in the absence of cell differentiation. Furthermore, 3T3-L1 adipocytes incubated with low concentration of CB-77 stimulated adipokine release whereas this effect was absent when cells were exposed to high CB-77 concentration. Overall, the assumption that the combined effects of different congeners are dose-additive, which is the most important prerequisite of the TEQ concept, is highly questionable and linear extrapolation procedure is likely to under-estimate the metabolic risks linked to POP exposure.

- TEF, TEQ and the assumption that dioxins and dl-PCBs exhibit similar types of biological effects through interaction with AhR. TEF concept assumes that, among POPs, TCDD is the most toxic chemical. Therefore, the potency of dioxins, furans and dl-PCBs has been expressed as fractions of the potency of TCDD. Although TCDD, through AhR activation, may induce cancer, reproduction and neuronal dysfunctions, TEQ is unlikely to predict the risk of metabolic disorders induced by POP exposure. For instance, the ability of different POP mixtures to impair insulin action in 3T3L1 adipocytes was reported to be independent of TEQ. Both PCDD and PCDF mixtures did not impair insulin-stimulated glucose uptake compared with vehicle-treated cells, whereas non- and mono-ortho-PCB mixtures induced significant disruption of insulin action. Interestingly, the stimulation of glucose uptake induced by insulin was highly impaired after exposure to mixtures of organochlorine pesticides. These in vitro findings are consistent with many human studies indicating that POPs increase the risk of diabetes independently of TEQ, and that some organochlorine pesticides, like trans-nonachlor, are more strongly associated with diabetes than other POPs. Consistently, in vitro experiments using single POP compound reported that, at similar TEQ value, CB-77 stimulated adipocyte differentiation whereas this effect was absent in cells treated with TCDD. In concert, these results suggest that the deleterious metabolic effects associated with POP exposure may occur independently of AhR activation. Potential modes of action of POPs may include activation of constitutive androstane receptor (CAR) or steroid xenobiotic receptor (SXR), and competitive binding to nuclear receptors.

- Adult and early life exposure: There is a special concern regarding children because regulatory policies, which established dioxins and dl-PCBs tolerable intake on a body weight basis, do not distinguish between children and adults. Because of their high food intake per kilogram body weight required to maintain whole-body homeostasis and growth, children are likely to be at higher risk for environmental pollutant exposure. Not surprisingly, many scientific studies have highlighted that children are over-exposed to dioxins and dl-PCBs, and exceed the TDI of 2 pg/kg body weight. Furthermore, both metabolism, absorption and excretion system are different in children compared with adults, so that this subpopulation may potentially be more responsive to environmental pollutants. Another important issue concerns foetal life. Generally, pregnant women have been advised to avoid the consumption of food containing elevated levels of POPs, as well as other environmental pollutants like heavy metals.

However, POPs bio-accumulate in the body for many years, and restricting the exposure to these pollutants only during pregnancy would not protect the foetus or the future generations against the harmful effects of these hazardous chemicals.

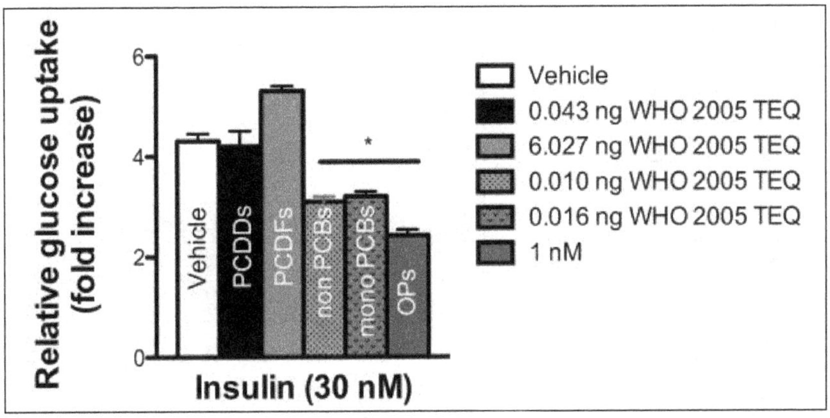

Effect of POP mixtures on insulin action: Differentiated 3T3-L1 adipocytes were incubated for 48 h with different POP mixtures mimicking those present in the oil of farmed Atlantic salmon. The ability of insulin (30nM) to stimulate glucose uptake was assessed. Note that the PCDF mixture and its elevated TEQ value did not impair insulin-stimulated glucose uptake. At the opposite, PCB mixtures and their low TEQ values significantly impaired insulin action. The most important inhibitory effect on insulin action was observed after exposure to organochlorine pesticides (1nM). PCDDs, polychlorinated dibenzo-p-dioxins; PCDFs, polychlorinated dibenzofurans; PCBs, polychlorinated biphenyls; OPs, organochlorine pesticides.

Inconsistency of POP Regulation in Animal Food Products

The omnipresence of POPs in animal food products has led international and national authorities to set up maximum limits. In this regard, marine food has attracted great interest and has been the source of intense debates. In the EU, maximum limit of dioxins, furans and dl-PCBs in fish and seafood products have been set to 8 pg WHO 1998 TEQ/g fresh weight. Giving that these POPs are highly lipophilic and therefore mainly present in the fat fraction of fish, this concentration could reach around 70 pg WHO 1998 TEQ/g fat for a fatty fish. On the other hand, the maximum limit of dioxins, furans and dl-PCBs in marine oil has been set to 10 pg WHO 1998 TEQ/g fat, a level that is seven times lower the one possibly observed in muscle meat of fatty fish. Thus, eating 1 g of fat from a fatty fish fillet could induce an exposure to 70 pg WHO 1998 TEQ (depending on the fat content of fish), whereas consuming 1 g of fat from a marine oil product could lead to an exposure to 10 pg WHO 1998 TEQ. Since the maximum levels for dioxins and dl-PCBs in foodstuffs should be set according to their potential harmful effects and risk to public health, one would expect these levels to be harmoniously regulated within food products.

Table: Regulation of dioxins, furans and dl-PCBs in fish and other animal food products by the EU.

	Maximum levels of dioxins, furans and dl-PCBs (WHO 1998 TEQ)
Meat and meat products from:	
Ruminants:	4.5 pg/g fat
Poultry and farmed game:	4.0 pg/g fat
Pigs:	1.5 pg/g fat
Muscle meat of fish and fishery products:	
	8.0 pg/g fresh weight (70.0 pg/g fat)[2]
Fat products from:	
Ruminants:	4.5 pg/g fat
Poultry and farmed game:	4.0 pg/g fat
Pigs:	1.5 pg/g fat
Marine oils:	10.0 pg/g fat

Another important issue is the regulation of organochlorine pesticides, which are chemicals strongly linked to type 2 diabetes as well as breast and prostate cancer and Parkinson disease. As for dioxins, furans and dl-PCBs, the EU has set up maximum residue levels in or on food and feed of plant and animal origin. Thus, fruits, vegetables, products of terrestrial animal origin, cereals, and other food items are regulated for their content in pesticides. However, levels of pesticides are still unregulated in fish and seafood in the EU. Similarly, there are still no maximum limits for PCBs in marine food. Because seafood represents one of the main sources of human exposure to POPs, and that POPs contribute to metabolic diseases, policy makers and stakeholders should urgently regulate and diminish the concentrations of POPs in seafood to protect the general population. Indeed, recent studies have reported an association between fish and seafood intake and type 2 diabetes in humans, and possible connection between farmed Atlantic salmon and the epidemic of type 2 diabetes has been suggested.

Further, some EU countries, like Finland and Sweden, have obtained dispensations to sell fish species from the Baltic sea, a highly polluted area where fish species can accumulate significant levels of POPs -dioxin levels higher than 4 pg/g fresh weight of muscle meat and fishery products are allowed. Also, the maximum levels of dioxins and dl-PCBs in muscle meat of eel, a potential highly contaminated fish, reached 12 pg WHO 1998 TEQ/g fresh weight, in contrast to the 8 pg WHO 1998 TEQ/g fresh weight of other fish species. Such derogations likely rose from the fact that background contamination of certain foodstuffs is already very high -mainly because of human pollution- and that keeping maximum dioxins and dl-PCBs levels similar in all food products would result to declare considerable part of our present food supply unfit for human consumption. Unfortunately, this strategy is not aimed to reduce human exposure to

POPs and raise serious questions about the effectiveness of EU regulations to protect consumers.

The global prevalence of metabolic diseases like obesity and type 2 diabetes, and its colossal economic and social costs represent a major public health issue for our societies. There is now solid evidence demonstrating the contribution of POPs, at environmental levels, to metabolic disorders. Thus, human exposure to POPs might have, for decades, been sufficient and enough to participate to the epidemics of obesity and type 2 diabetes. Based on recent studies, the fundaments of current risk assessment of POPs, like "concept of additive effects" or "dioxins and dl-PCBs induced similar biological effects through AhR", appear unlikely to predict the risk of metabolic diseases. Furthermore, POP regulation in food products should be harmonized and re-evaluated to better protect consumers. Neglecting the novel and emerging knowledge about the link between POPs and metabolic diseases will have significant health impacts for the general population and the next generations.

OBESITY

Non-communicable diseases have considerably increased worldwide during the last decades. The increase in obesity prevalence is particularly relevant since it is a commonly known risk factor for disorders such as impaired glucose tolerance, metabolic syndrome, diabetes mellitus, liver and cardiovascular diseases (CVD), as well as some cancers. The adipose tissue (AT) of obese individuals is quantitatively much larger and includes more pathological features, than that of lean individuals. Much of our understanding of the interaction between obesity and environmental pollutants is largely focused on the AT. Historically, the AT was considered as a simple storage tissue. However, its physiological functions have been considerably reassessed over the last decade. Evidence for metabolic, endocrine and immune functions of the AT including stroma has accumulated. Greater attention is now given to the pathological contribution of the AT to obesity and metabolic disorders such as type 2 diabetes. Lately, various interactions between the AT and certain pollutants such as the persistent organic pollutants (POPs) have been established suggesting that this tissue plays a significant role in the kinetics and the toxicity of POPs.

POPs cannot be metabolized by the xenobiotic metabolizing system and therefore tend to accumulate in ecosystems and in living organisms. The best studied are those which were listed in the Stockholm convention to limit their production and dissemination because of their possible long term toxicity. POPs include certain organochlorine pesticides, dioxins, furans, polychlorobiphenyls and polybrominated flame retardants. They do not readily undergo degradation by xenobiotic metabolizing enzymes (XMEs), because of their bulk and halogenation. However, they do activate certain xenobiotic receptors, and some bind to certain XMEs such as CYP1A2 without undergoing catalytic transformation. Because of their hydrophobicity, POPs tend to distribute into lipid rich tissues such as the AT and milk.

The Obesogen Concept

Exposure to certain pollutants during particular windows of vulnerability has been shown to increase AT mass and contribute to obesity later in life. Development, e.g. prenatal, postnatal, and pubertal, is likely a critical window of susceptibility to obesogen effects of toxic exposures. Programming mechanisms are still unclear, but are believed to involve epigenetic regulation of critical genes that lead to adiposity later in life. Evidence suggests that developmental exposures to chemicals that increase risk of obesity sometimes operate in a non-monotonic dose-response manner; cachexia may occur at high doses whereas body and/or adipose mass gain occurs at low doses of the same chemical. Further, there may be gender specific effects of developmental toxic exposures that increase the risk of obesity. Developmental exposures to these same POPs are positively associated with obesity in humans.

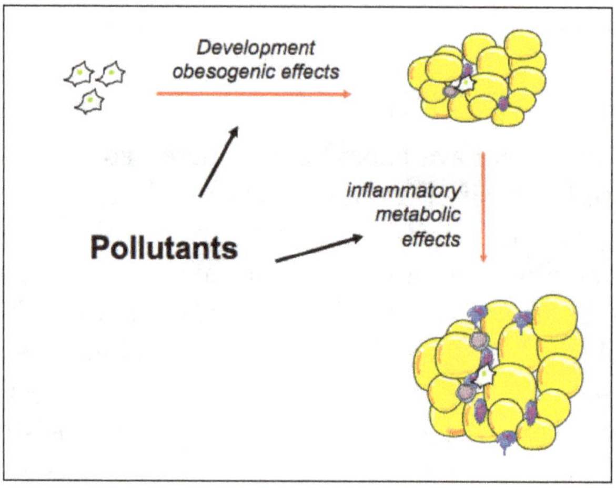

Model representing the effect of pollutants on AT: Exposure to several chemicals called obesogens during the perinatal period leads to the development of obesity later in life. The mechanisms of this programing effect have not been delineated, however, it is believed that epigenetic regulations are involved. Pollutants, particularly POPs, can also interfere with AT biology either by increasing inflammation or through metabolic disruption and thereby contribute to the appearance of pathological side effects of obesity. Such mechanisms may also take place at the adult stage and are therefore distinct from the obesogen effects.

Obesogens are frequently endocrine disruptors and belong to several chemical families. Several studies have been carried on POPs which are either dioxin-like (DL), ie they mimick the effect of dioxin on the dioxin receptor, AhR, or non DL. Rodent models indicate that DL chemicals may be obesogens. Exposure to TCDD (2,3,7,8-tetrachlorodibenzo-p-dioxin), 100 µg /kg b.w. once every 2 weeks for 8 weeks, increased body weights of adult mice over 40% higher than control-treated C3H/HeN mice. This body weight change was only seen when mice were fed a high fat diet. In a one-month study, chronic developmental exposure to the PCB mixture Aroclor 1254 was associated with increased body weights of mouse pups on postnatal days (PND) 16-20. Further, exposure of adult mice PCB-77 led to an AhR dependent increase in body mass. This PCB-77 exposure also increased fatty liver in CVD model mice. Fatty liver, attributed to increased hepatic triglycerides and cholesterol, was also caused by 50 mg of PCB-169/kg body weight.

There is limited evidence of increased adiposity in animal studies of POPs that are not DL, however body fat is seldom assessed in studies reporting no increased body mass after POP exposure. Prenatal exposure to a major polybrominated- diphenyl ether (PBDE-99, 2,2',4,4',5-penta-BDE) increased mouse birth weight, and pre- and postnatal exposure to BDE-47 (2,2',4,4'-tetra-BDE) increased rat body weights from birth to puberty (when the study ended). In the longest study of developmental PBDE exposure to examine body weights, male mice exposed to BDE47 10 days after birth had increased body weights from PND 47 until the end of the study, at 4 months of age. These studies indicate significant body composition effects of perinatal exposure to PBDEs, however the mechanisms remain unclear and the data should be interpreted with caution as certain preparations of BDEs could be contaminated with DL chemicals. In perinatal exposure to perfluorooctanoic acid (PFOA) which is not a traditional POP, obesogenic effects do not appear until later in life. Mice exposed to low levels of PFOA in utero had increased body mass once mature, with an inverted U shape dose response curve. By 18 months of age, there was no longer an effect on mouse weight, however, there was a positive dose response relationship between in utero PFOA exposure levels and abdominal brown AT mass in the aged mice, whereas a negative relationship was found with white AT mass. Consistent with experimental findings, a recent prospective human study demonstrated that maternal PFOA levels during pregnancy were associated with obesity in the daughters 20 years later. Organochlorine pesticides may also increase adiposity. For instance, oral DDT exposure increased the body weights and/or adiposity of both mouse and rat offspring in several multi-generational studies.

Several studies assessing obesogenic effects were devoted to other pollutants which are not persistent, particularly endocrine disruptors. Much effort was devoted to Bisphenol A (BPA), and, in many cases, perinatal exposure to BPA was shown to lead to increased body weight later in life. Non monotonic curves were sometimes observed. However, the effect of BPA was not always consistent, suggesting that specific experimental conditions were required to unravel the obesogenic effects of BPA. This could be interpreted as suggesting that the interaction of this chemical with other environmental factors such as diet is critical for the obesogenic effect to be observed. Human studies recently demonstrated that fetal exposure to BPA was associated with increased BMI in 4 years old children. Another well studied obesogen is tributyltin (TBT). Perinatal exposure to TBT leads to increased adipose mass transgenerationally. These effects have been related to the activation by TBT of the PPARg/RXRa heterodimer and with a possible involvement of epigenetic effects. The effects of more complex exposures such as maternal smoking and air pollution, which also correlate with offspring obesity.

Are Metabolic Consequences of Obesity Induced by Pollutants?

Obesity causes predisposition for other metabolic diseases such as type 2 diabetes and and metabolic risk features such as moderate elevation of glycemia, hypertriglyceridemia or low HDL. Several epidemiological studies carried following industrial exposure of workers or accidental contamination by POPs indicated a relationship between serum concentration of certain POPs and markers of diabetes or of a prediabetic state. This is the case of the Seveso cohort in which increased metabolic syndrome (but not obesity) was observed in women exposed to dioxin before the age of 12. Such a correlation was also found in a large study carried in the general population. Prospective studies in the elderly have also indicated a possible role of certain POPs in the pathogenesis of type 2 diabetes. There are also some experimental studies clearly correlating POP mixtures with metabolic syndrome occurrence in the rat. A National Toxicology Program workshop concluded that POPs were associated with type 2 diabetes but that no causal relationship could be established at this stage. A recent study indicated that metabolically healthy but obese individuals had lower plasma levels of several classes of POPs than obese individuals with metabolic abnormalities.

The interactions between obesity, POPs and metabolic disruption were unraveled in several mechanistic studies. Because of the implication of the AT in metabolic diseases, it was hypothesized that this tissue could be a target of POPs and indeed, several effects were found. Its vulnerability may be due to its ability to accumulate POPs. Most of the studies were in vitro or ex vivo, but recently the effect of POPs on the AT of rodents was also assessed. POPs were shown to display anti-insulin effects in cellular models of adipocytes. For example, dioxin repressed the glucose transporter Glut4 expression and lipoprotein lipase in 3T3-F442a cells. This anti-insulin effect is not general and consistent for all genes. Indeed, whereas dioxin was found to antagonize insulin action on certain genes such as the IGFBP1 gene in hepatocytes, it displayed a different effect on other genes such as the liver PEPCK gene, since it tended to inhibit gluconeogenesis in this tissue, similarly to insulin.

Inflammation of the AT is one of the hallmarks of obesity and inflammatory phenotype is critical in metabolic diseases. POPs have been shown to induce proinflammatory genes in rodent adipose cells. Importantly, in mice treated with dioxin, not only the gene expression of proinflammatory genes was increased, but also invasion of this tissue by macrophages and lymphocytes was observed. Finally, dioxin was shown to inhibit the differentiation of adipocyte precursor cells in certain model systems and to antagonize the effects of PPARg. However the actual mechanisms remain elusive. In conclusion, preadipocytes and adipocytes are targeted by POPs which appear to disrupt certain signaling and differentiation pathways and to induce inflammation.

Is there a Protective Role of AT and Obesity?

The AT is a compartment which contains a high amount of POPs, particularly in organisms that are at the top of the food chain. Such a bioaccumulation leads to the age-dependent increase in POP content. POPs are taken up by adipocytes and localize within lipid droplets. However their precise location and their actual effects at the subcellular level are poorly understood. It is nevertheless believed that their accumulation within the AT decreases their availability for other cells and tissues thereby limiting their toxicity. Experimental evidence supports such a protective function for the AT. Indeed, studies carried in the 80s and the 90s showed that there was an inverse correlation between toxicity of POPs and fat mass of different animal species. Authors compared the 30-day toxicity of TCDD in approximately twenty terrestrial animal species and found a positive correlation between the BMI of these species and the LD50 (the dose that leads to 50% death in the animal population) of dioxin. They concluded that the species with the highest fat mass tended to display better resistance to dioxin in this particular acute exposure test. These conclusions were in line with studies showing that resistance of aquatic species to dioxin was also related to their fat mass leading to the paradoxical notion of "survival of the fattest". However, these observations should not be taken as evidence suggesting that the BMI is the only factor discriminating sensitive and resistant species. There is indeed strong evidence for a major contribution of the genetically determined arylhydrocarbon receptor affinity for dioxin.

It should be stressed that this protective function of the AT was revealed in acute or subacute exposure tests. These high dose treatments may allow the distribution of the pollutants to all tissues unless an efficient "filter" or a buffer system can capture them, thereby decreasing exposure of the most sensitive tissues. This role is played by the partitioning of POPs into lipid-rich tissues. This kinetic protective system does not only include the AT. Indeed, it has been established that proteins such as the dioxin-inducible liver CYP1A2 can bind this pollutant particularly during acute or subacute exposures and play an important role in its toxicokinetics. It is now believed that POPs are first distributed throughout the body and then captured by the liver inducible protein compartment, with excess then redistributed to the AT. Obviously, these kinetic distribution mechanisms depend heavily on the treatment dose and the body burden. Furthermore, in several metabolic disorders, lipid droplets are found in other tissues such as liver, muscle, heart, etc. with possible consequences related to POP storage.

AT could also be involved in the higher blood half-life of POPs in children. Indeed, newborns have a high body burden due to placental transfer during pregnancy and to breast-feeding. A higher blood elimination rate observed for the children compared to the adults might be explained by the dilution of the POPs across tissues like the AT rather than a higher metabolic rate.

There is some indirect evidence for a protective role of AT from human studies. The association between fat mass and mortality depends on the serum concentration of POPs.

Indeed, in those individuals with low POP concentration, mortality increased with fat mass, whereas in those with high POP concentrations, mortality tended to paradoxically decrease with fat mass. These observations can be accounted for by a protective function of the AT which becomes significant at high levels of POP contamination.

Is the AT a Source of Endogenous Exposure?

As mentioned earlier, POPs and other lipophilic contaminants distribute according to their affinity for proteins and lipids and are stored primarily in the liver and the AT. They are also found in blood from which they can contaminate other tissues. Blood POP content can be either related to their release from storage tissues or to recently absorbed pollutants. Several observations in both human and animals suggest that the release of pollutants from the AT is an important source of blood POPs.

In humans, most of the evidence has been gathered from studies on drastic weight loss in obese individuals. Such a weight loss can be achieved voluntarily through diet and bariatric surgery and could lead to a decrease of up to 30 kg of fat mass or even more in some cases. Several independent studies have shown that there was an increase in blood POPs following fat mass loss elicited by either diet alone or diet coupled with bariatric surgery. If increased blood POP levels during weight loss is related to their release from AT, one would expect changes in POP content of this tissue. This has been addressed by Kim et al who determined POP concentrations in both blood and AT and who also assessed the total amount of fat in the studied individuals. The data indicate that POP concentration in AT (expressed per gram lipid) increases with weight loss. While this may seem paradoxical, it is not particularly surprising since the total amount of fat mass decreases considerably thereby leading to an increased concentration of pollutants, e.g. released POPs can be taken up readily by the remaining fat. In line with these suggestions, we observed that POP concentrations in the AT of obese individuals is lower that of lean individuals. However, the total amount of fat-stored POPs is 2- to 3-fold higher in obese individuals as compared to lean controls. Furthermore, this total amount tends to decrease by 15% following weight loss at least for certain POPs. This observation suggests that there is indeed some degree of POP release from AT during weight loss and that this release leads to a moderate decrease in total POP content.

Experimental evidence also suggests redistribution of POPs from their storage sites in the AT. Indeed, a study shows that in rodents pretreated with radiolabeled hexachlorobenzene, weight loss leads to a time-dependent increase in the brain content of this compound. The study shows that weight loss alters the distribution of lipophilic pollutants, thus leading to enhanced localization in the brain and other sensitive tissues with possible toxic outcome.

Observational studies were also carried in northern elephant seals. These animals accumulate a large amount of fat in order to cope with extended fasting. Their fat is contaminated with PCBs. During the fasting period which could last several weeks, they lose a large amount of fat. An increase in serum concentration of PCB during fasting which is likely due to their release from fat depots. Interestingly, the concentration of PCBs also increased in some of these depots (blubbers) because of the decreased fat content. However, different fat depots did not undergo similar changes, suggesting differences in the kinetics of POP exchange and release. It is suggested that the release of POPs during fasting may lead to toxic effects.

A critical issue is whether the release of POPs from AT observed during weight loss could lead to toxic outcomes in other organs and tissues. Indirect evidence was obtained in humans from several studies of weight loss triggered by either diet or diet associated with bariatric surgery. We have shown that the dynamic increase in serum POPs following drastic weight loss correlated with a delayed and reduced improvement of blood lipid parameters and liver toxicity biomarkers. Correlations between blood POP concentrations and other clinical parameters such as metabolic and muscle parameters, were also observed in humans by the group of Tremblay who conducted seminal studies in this field.

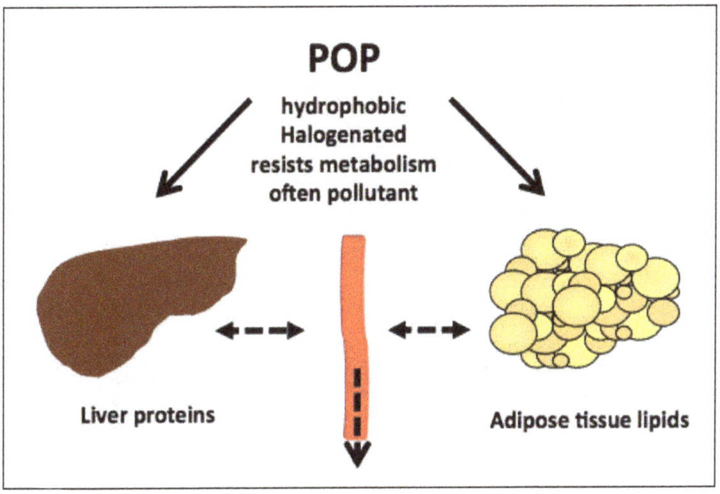

Fate of a persistent organic pollutant (POP): Most xenobiotics are metabolized primarily by the liver and are thus detoxified. The detoxification system tends to render hydrophobic xenobiotics more hydrophilic which leads to their elimination in urine. Several halogenated xenobiotics are not metabolized and therefore tend to bind to liver proteins and to adipose mass. They can thus persist for years in the body and constitute a putative long term threat since they can be released from these compartment at low levels.

The AT appears to play critical roles in the kinetics of POPs and in their pathogenic effects. It has a major role, together with the liver protein compartment in storing POPs and in preventing their distribution into more sensitive tissues. However, the AT storing capacity is constitutive and not inducible. This kinetic system acts as a buffer during acute or subacute exposure conditions. However, it translates an acute exposure into a long term, low-grade internal exposure. It thus transforms an immediate threat into a latent chronic threat. This buffer system perfectly illustrates a previously developed hypothesis which proposes that systems that protect from acute exposure to xenobiotics contribute to their chronic toxicity. In addition to these functions, the AT constitutes a target of POP toxicity. Indeed, the main toxic effect triggered by these compounds is inflammation which is a well-known risk factor for metabolic diseases. These observations support the contribution of POPs to metabolic diseases and suggest that AT alteration could at least partially mediate these effects.

Remediation and Prevention

6 CHAPTER

The restoration of contaminated environmental sites in such a way that it is no longer a threat to human health is called remediation. There are various bioremediation and prevention technologies used such as bioventing, land farming, bioreactor, composting, bioaugmentation and biostimulation. This chapter has been carefully written to provide an easy understanding of the subject of remediation and prevention of organic pollutants.

REMEDIATION OF POP CONTAMINATED ENVIRONMENTS

Remediation refers to the restoration of contaminated environmental sites to a condition that is not a threat to human health or other lifeforms. In most cases, a great deal of effort, human and technical resources are used to clean up contaminated sites. In many cases, sites are so heavily polluted, that they can only be fenced off and isolated as much as possible from the rest of the environment. During the last decades new products and technologies have been developed for remediating ground water, surface water, soils, air and sediments. Technologies that are used to remediate sites are also frequently used to treat industrial effluents before they are released into the environment. These waste treatment and remediation technologies and products can include incineration, activated carbon absorption, ion exchange, soil washing, chemical precipitation, oxidation, encapsulation of wastes into cement and other binders, electrolytic treatment, elektrokinetic remediation, "in situ vitrification" (melting soils in place which leads to the incineration or encapsulation of the contaminants), or biological treatment with plants, fungi, and bacteria (this is called "bioremediation"). For remediation of POP polluted sites bioremediation is particularly useful.

Bioremediation

Generally, bioremediation technologies can be classified as in situ or ex situ. In situ bioremediation involves treating the contaminated material at the site while ex situ involves the removal of the contaminated material to be treated elsewhere. Some examples of bioremediation technologies are bioventing, land farming, bioreactor, composting, bioaugmentation and biostimulation.

There are only few reports available on the degradation of POPs. HCB and dieldrin are still found in the environment such as agricultural or/and paddy field even more than 30 years since their prohibition. Thus, contamination with organochlorine pesticides is still a serious environmental problem which requires effective and efficient methods for remediation. Among the removal technologies for such residual pesticides, bioremediation is considered to be the most cost-effective because residues of HCB, dieldrin or s-triazines are known to have low concentration and a wide range. Microbial degradation is a promising effective technique to remediate environmental pollutants. For HCB, several studies have reported reductive dechlorination and conversion of HCB to pentachlorophenol (PCP) by using genetically engineered bacterial cells.

The microbial degradation of soil pollutant POPs is a fast growing field. Novel bioremediation technologies were developed for POP-contaminated soil and water using an aerobic pentachloronitrobenzene (PCNB)-degrading bacterium, Nocardioides sp. Strain PD653, which was isolated from an enrichment culture in a soil-charcoal perfusion system. Not only bacteria, but even certain fungi can degrade POPs, such as the aerobic dieldrin-degrading fungus, Mucor racemosus strain DDF, which was isolated from annually applied soil with endosulfan. DDF was able to degrade dieldrin more than 90% for 10 days incubation whereas previously reports degraded dieldrin less than 50%. However, the application technologies to the contaminated sites are still inadequate to remediate s-triazine-contaminated sites. Therefore, Takagi and co-workers developed a new method to introduce the degrading-bacterial consortium into contaminated soil using a special charcoal material which was enriched with a methylthio-s-triazine degrading bacterium and the chloro-s-triazine degrading bacterial consortium CD7. The enriched charcoal was capable of degrading chloro-s-triazines (simazine and atrazine) and methylthio-s-triazines (simetryn and dimethametryn) simultaneously in sulfur-free medium. Moreover, using enriched charcoal with CD7 in situ bioremediation study was conducted in a contaminated site, where simazine is routinely applied for preservation of its turf. The material was effective for preventing penetration of simazine into subsoils and aquatic environments nearby for approximately 2 years. For microbial POP removal technologies genetic engineering is the most important tool, but it may raise legal issues regarding the restrictions against GMOs particularly in case of open field in situ applications. The GMO problem might become even more obvious, when higher plants are engineered for remediation of POPs, in this case the term phytoremediation is preferred. However, the great advantages of bioremediation/phytoremediation should be considered against the risks of application of genetically engineered plants particularly unwanted gene flow to the environment, these plants applied with necessary caution after a thorugh environmental risk assessment, will be able to solve a number of serious environmental problems caused by POPs.

However, even non-GMO alternatives can work efficiently, such as tobacco plants and the paraquat resistant poplar, which have been developed by traditional clone selection methods. Poplar (Populus ssp) trees are good candidates for phytoremediation purposes due to the extensive root system, high water uptake, rapid growth, and large biomass production.

Poplars have already been used to remove atrazine (2-chloro-4- (ethylamine)-6-(isopropylamine)-s-triazine), trichloroethylene and selenium from polluted soils. In in vitro studies, poplar also showed an extremely high capacity for heavy metal uptake coupled with elevated genes expressions (gsh1) and enzyme (GST) activities. Paraquat (1,1-dimethyl-4,4-bipyridinium dichloride) is a non-selective contact herbicide that targets the plant chloroplasts, causing continuous oxidative stress, membrane damage, chlorophyll loss and desiccation. Bittsánszky and coworkers selected in vitro nodal segment cultures at the sublethal paraquat concentration (0.4 μmol L−1). After testing on tissue culture media, regenerants were micropropagated, rooted and transplanted to a greenhouse followed by a field test at heavily contaminated fields. They found, that Paraquat-tolerant clones of poplar were powerful new stress tolerant and non-GMO plants, which can provide an effective tool for phytoremediaton purposes where the pollutants cause oxidative stresses.

Electrokinetic Remediation (EK)

Electrokinetic remediation (EK) and its combined methods are also emerging technologies developed during the last decade in order to remove POPs from contaminated soils. Electrokinetic remediation is an in-situ process in which an electrical field is created in a soil matrix by applying a low-voltage direct current (DC) to electrodes placed in the soil. As a result of the application of this electric field, not only heavy metal contaminants, but even POPs may be mobilized, concentrated at the electrodes, and extracted from the soil. Interest in electrokinetic remediation has been driven by the demand for technologies that are cost effective and will eliminate the long-term liability that is incurred by landfilling of contaminants. There are five major methods of electrokinetic remediation: (1) Classic electrokinetic method; (2) Electrokinetic-Fenton method; (3) Electrokinetic remediation combining with bioremediation method; (4) Electrokinetic remediation combining with surfactants/cosolvents method; (5) Electrokinetic remediation combining with ultrasonic method. The intensive future development of EK remediation for POP contaminated soils can be anticipated, even if this method is not as cost effective as bioremediation.

NUTRITIONAL INTERVENTION

Environmental pollution is increasing worldwide, and there is evidence that exposure to halogenated persistent organic pollutants (POPs) such as polychlorinated biphenyls can contribute to the pathology of inflammatory diseases such as atherosclerosis, diabetes, and cancer. Pollutant removal from contaminated sites and subsequent pollutant degradation are critical for reducing the long-term health risks associated with exposure. However, complete remediation of a toxicant from the environment is very difficult and cost-prohibitive. Furthermore, remediation technologies often result in the generation of secondary toxicants. Considering these circumstances, environmentally-friendly and sustainable remediation technologies and biomedical solutions to reduce vulnerability to environmental chemical insults need to be explored to reduce the overall health risks associate with exposure to environmental pollutants. Positive lifestyle changes such as healthful nutrition and consumption of diets rich in fruits and vegetables or bioactive nutrients with antioxidant and/or anti-inflammatory properties will reduce the body's vulnerability to environmental stressors and thus reduce toxicant-mediated disease pathologies. Interestingly, emerging evidence now implicates the incorporation of bioactive nutrients, such as plant-derived polyphenols, in technologies focused on the capture, sensing and remediation of halogenated POPs. Human nutritional intervention in concert with the use of natural polyphenol sensing and remediation platforms may provide a sensible means to develop primary and long-term prevention strategies of diseases associated with many environmental toxic insults including halogenated POPs.

Combining Environmental Science and Biomedical Strategies to Better Protect Against POP Toxicity

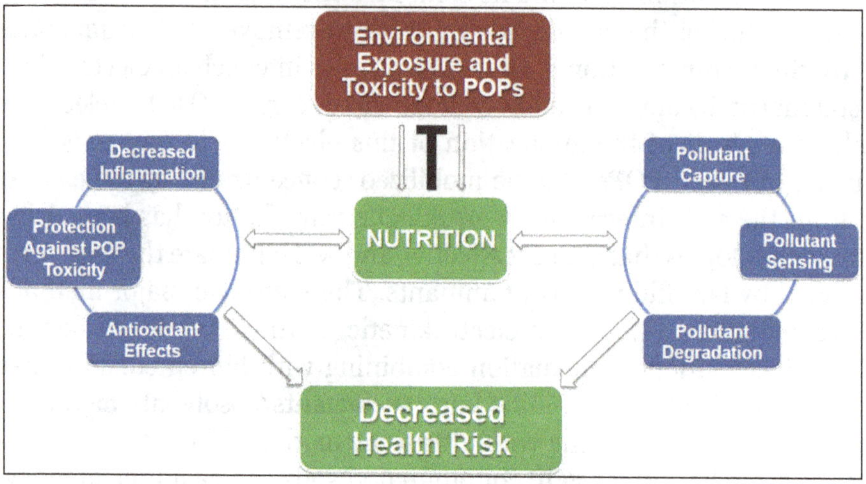

Contaminant remediation and biological modulation of environmentally persistent pollutants are two crucial means of reducing the human health risks of Superfund chemicals/toxicants and related hazardous materials. The Superfund program, designed to clean up sites contaminated with a variety of hazardous substances, was formed as the United States' response to this growing environmental and public health concern, but these pollutants are of worldwide concern. Although complete remediation of a hazardous site may continue to be an ultimate goal of human exposure risk reduction, it is important to design and implement other biologically relevant means of buffering against toxicant exposure before, during, and after remediation activities. New data now implicate the importance of an individual's nutritional status and the use of protective bioactive food components to decrease the overall toxicity of environmental pollutants to biological systems. Nutrition is being substantiated as an important modulator of inflammatory and antioxidant pathways, especially with regard to environmental insults. Interestingly, novel research now shows that bioactive food components that are environmentally friendly can also be integrated into remediation technologies, which in turn allow for more sustainable, inexpensive, and effective pollutant removal and detoxification. Implementing bioactive molecules in both biomedical and environmental science settings, together, will allow for decreased overall body burden and human toxicity of a multitude of pollutants.

Demonstration of nutrition as: 1) a modulator of health risks associated with environmental exposure and toxicity of persistent organic pollutants (POPs) and as 2) an environmentally sustainable tool to capture, sense and remediate POPs. Nutrition, and in particular diet-derived bioactive compounds such as polyphenols, can protect against POP toxicity through their antioxidant and anti-inflammatory properties. Polyphenols also can participate in a remediation platform that includes their roles in POP capture and sensing technologies, thus allowing for environmentally friendly and sustainable remediation of POPs.

Exposure to Pops, Risks for Non-communicable Diseases and Mechanisms of Toxicity

Halogenated persistent organic pollutants (POPs), including polychlorinated biphenyls (PCBs) and trichloroethylene, are highly prevalent in the environment in soil, air, and water. For example, a major source of human exposure to PCBs is through dietary intake of contaminated foods and through inhalation of airborne pollutants. Because most halogenated POPs, including PCBs, are lipid soluble, they easily accumulate in human tissues, leading to a perpetually increasing disease risk throughout a life span, especially in overweight populations. Exposure to halogenated POPs, and in particular pollutants that are ligands of the aryl hydrocarbon receptor (AhR), can lead to inflammatory events tied to diseases such as atherosclerosis, diabetes, obesity, cancer, etc.. The pathology of many inflammatory diseases, especially of non-communicable or chronic diseases, develops over a long period of time and thus can be easily modulated by environmental exposures, specifically to persistent organic pollutants.

Mechanistically, an increase in cellular oxidative stress often precedes an inflammatory response. PCBs, and in particular coplanar PCBs, have been shown to cause oxidative stress primarily through a cytochrome P450 (CYP1A1)-mediated uncoupling mechanism. CYP1A1 induction allows for the detoxification of multiple xenobiotics, but when in the presence of PCB, can become inefficient and leaky (i.e., uncoupled) and produce detrimental reactive oxygen species. A hallmark of the pathology of vascular diseases, including atherosclerosis, includes a change in the cellular redox status and a resultant increase in oxidative stress, which favors chronic and low level inflammation. Such changes in redox status and oxidative stress levels can be driven in part by pro-oxidative and pro-inflammatory environmental pollutants that are persistent and which can be easily stored in adipose tissue. Thus, an increased body burden of persistent environmental pollutants is a particular risk factor during obesity, a disease characterized by excessive adipose tissue. There is sufficient evidence that POPs contribute to inflammation by activating oxidative stress-sensitive transcription factors such as nuclear factor kappa-light-chain-enhancer of activated B cells (NFκB). For example, our studies suggest that PCBs, and in particular coplanar PCBs, can increase cellular oxidative stress and induce inflammatory parameters such as inflammatory cytokines, chemokines, and adhesion molecules in the vascular endothelium, which are metabolic events that foster an inflammatory response and atherosclerosis. Through these pro-inflammatory mechanisms, PCBs and related environmental toxicants have been correlated with increased risk of multiple human chronic disease phenotypes including diabetes and heart disease. Since many populations susceptible to toxicant-induced disease are often also afflicted by diet-induced diseases, future human studies and integrated risk assessments should better investigate the interaction between nutrition and toxicology.

Nutrition as a Modulator of Environmental Chemical Insults

Sub-Optimal or Unhealthy Nutrition Modulates the Toxicity of Environmental Pollutants

POPs and subprime nutritional status share mechanisms of disease development including the induction of pro-inflammatory pathways. Interestingly, a growing body of evidence implicates an exacerbated toxic effect of PCBs and other environmental pollutants when combined with an additive pro-inflammatory dietary environment. For example, when combined, pro-inflammatory PCBs and omega-6 fatty acids create an exacerbated toxicological response that is not simply the sum of their individual inflammatory responses. Importantly, animal fats such as those found in beef and chicken are significant sources of pro-inflammatory medium and long-chain omega-6 fatty acids as well as PCBs and other related POPs. Also, other groups have investigated a linkage between high saturated fat diets and increased toxicity of POPs such as polycyclic aromatic hydrocarbons and have shown an increased risk for adenomas and cancer. In addition, other environmental contaminants such as heavy metals may work in concert with sub-optimal nutrition to promote chronic inflammation and disease, but studies that investigate the impacts of mixing multiple types of exposures, (e.g. chlorinated pollutants and heavy metals) in combination with the added variable of an unhealthy diet are severely lacking. These types of comprehensive integrated toxicity studies will better mirror real world exposure conditions especially for people residing in close proximity to Superfund and other hazardous waste sites. Although obesogen research (e.g. toxicants promoting obesity) is becoming a more developed area of study, work is still needed to elucidate interactions between nutrients, nutritional status, environmental toxicants and human health.

Healthful Nutrition Decreases the Toxicity of Pro-Inflammatory Pollutants

Polychlorinated biphenyls and related Superfund POPs induce chronic oxidative stress and disregulated inflammatory responses, but anti-inflammatory nutritional antioxidants may buffer and protect against toxicant-induced disease through multiple cell signaling mechanisms. Polyphenols and omega-3 polyunsaturated fatty acids have been shown to decrease toxicant-induced maladies including liver diseases, tumor formation and growth and endothelial cell activation. Plant-derived flavonoids such as epigallocatechin-3-gallate (EGCG), and long-chain omega-3 fatty acids such as docosahexaenoic acid (DHA) can protect cellular systems by decreasing pro-inflammatory lipid raft signaling domains called caveolae and by simultaneously upregulating antioxidant defenses through increased nuclear factor (erythroid-derived 2)-like 2 (Nrf2) activation. Most recently we have elucidated novel cross-talk mechanisms between caveolae and Nrf2 pathways and have shown that nutrients and/or bioactive food compounds may protect against vascular dysfunction, oxidative stress and inflammation by downregualting caveolae and simultaneously upregulating Nrf2 target antioxidant genes. Bioactive food compounds such as those found in broccoli (sulphoraphane) and red wine (resveratrol) can activate Nrf2 through multiple mechanisms and decrease oxidative stress levels in cells and in animals. Polyunsaturated fatty acids, and especially omega-3 fatty acids, as well as their anti-inflammatory breakdown metabolites can protect against PCB-induced disease by activating Nrf2 but interestingly also by disrupting functional caveolae. Although caveolae and Nrf2 pathways appear to be prime candidates for bioactive food components to work through, more research is necessary to better elucidate other novel pathways that can be targeted both *in vitro* and *in vivo*.

Healthful Nutrition can Decrease Body Burden of Environmental Toxicants

An emerging paradigm implicates healthful nutrition as an effective protective modulator of environmental toxicant-induced inflammation and human disease. Multiple laboratories have investigated the decreased toxicities of environmental pollutants due to bioactive nutrients such as flavonoids and omega-3 polyunsaturated fatty acids, but many of these studies rely on in vitro assays that lack the complexity of a whole body organismal approach. Importantly, emerging classes of bioactive food components such as polyphenols also have been shown to modulate the pro-inflammatory effects of environmental toxicants. Array of phenolic compounds such as EGCG, curcumin and quercetin can decrease toxicant-induced oxidative stress and inflammation in multiple cell types, tissues and animal species Although human studies are lacking, strong evidence in animal models implicates a protective role for flavonoids and other polyphenols perhaps through the induction of antioxidant enzyme pathways and increased fecal excretion rates. Since the protective effects of healthful nutrition against biological insults induced by exposure to POPs, the remainder of this review will focus on what is known concerning the overall decrease in body burden of POPs, with a special emphasis on PCBs and related compounds, by nutrients such as polyphenols. In other words, plant-derived polyphenols are being found to not only protect against POP-mediated oxidative stress, inflammation, and toxicity but also to bind to POPs and thus contribute to a decrease in body burden. In fact, if has been proposed that green tea, containing high levels of polyphenols including EGCG, can inhibit the intestinal absorption of lipids and highly lipophilic organic compounds and accelerate excretion of PCBs.

Certain foods and polluted air can be major sources of toxicant exposures. Thus, altering nutritional choices may prove to effectively modulate risks associated with exposure to POPs. Making informed dietary decisions such as substituting lower fat versions of protein sources (e.g. legumes, nuts, or lean meats and dairies) may have multiple health benefits including decreased exposure to detrimental pollutants such as dioxins and PCBs. For many populations it may be difficult or cost-prohibitive to change major dietary protein sources; thus, increasing dietary intake of other bioactive nutrients may buffer already exposed individuals against the negative ramifications of pollutant exposures. For example, several laboratories have shown that diets high in fiber can alter the absorption and excretion rates of pollutants such as PCBs.

Mechanistically, multiple dietary fibers have been shown to effectively bind pollutants such as dioxins, which may help to explain increased fecal excretion rates, decrease in body burden, and observed protection. Increasing excretion of lipophilic toxicants through other nutritional compounds such as fat substitutes may also effectively decrease risk. For example, much emphasis including human clinical trials has been placed on the interactions between the non-absorbable fat substitute olestra and environmental toxicants. Researchers determined that low levels of olestra supplementation (25 grams/day) increased the excretion of multiple PCBs and related contaminants upwards of 11 fold compared to normal diet. Interestingly, in a human case study, an olestra-supplemented diet completely eliminated excessive POP concentrations in adipose tissue and reversed POP-mediated clinical conditions such as diabetes and hyperlipidemia. Diet-derived bioactive compounds such as EGCG are attractive modulators of toxic exposure because they may both help to prime the physiological system prior to a toxic insult by upregulating protective detoxifying enzymes as well as protect therapeutically after exposure by increasing the rate of excretion and lowering overall body burden. Altering diets to increase fiber and bioactive nutrients from fruits and vegetables is an effective and cost-efficient means of modulating POP toxicity, but interestingly an emerging and growing body of knowledge also implicates the use of bioactive nutrients such as quercetin and curcumin as effective components of novel environmentally friendly remediation technologies.

Environmental remediation not only relies on efficient pollutant degradation but initially on effective pollutant sensing and capture/removal from contaminated settings. Emerging research suggests that the utilization of polyphenols for capture and sensing of POPs may provide a technologically efficient and cost effective remediation platform that results in minimal environmental and health impacts. This work adds to the paradigm that nutrients or diet-derived bioactive compounds can broadly impact health concerns associated with POP exposure.

Utilizing Polyphenols for Pollutant Sensing, Binding and Remediation Efforts

Persistent organics, and especially PCBs, have proven difficult to remove from contaminated sources and subsequently remediate due to their poor aqueous solubility and low volatility, which makes extraction from soil and water especially challenging. Further, sensing of these pollutants is both difficult and cost prohibitive, relying heavily on GC-MS methods and complicated extraction techniques. POP remediation to date often involves dredging and subsequent deposition in landfills or further pollutant degradation through incineration or various chemical dehalogenation techniques. Studies have shown, though, that dredging and deposition results in substantial leaching into the surrounding environment, incineration can produce even more harmful byproducts if insufficient temperatures are reached, and organic solvents used for chemical dehalogenation are often as toxic as the pollutants being remediated. There is a clear need for more effective pollutant remediation involving inexpensive, environmentally friendly, stable systems for the capture/sensing of PCBs and other POPs, as well as environmentally-conscious pollutant degradation. In fact, as of 2011, the Ecologic Institute of the European Union's Screening methods for the Water Framework Directive (SWIFT-WFD) has continued to focus on this need for discovery and translation of low-cost screening tools for environmental contaminants.

As a result, much research has been performed to determine novel technologies and methods to capture, sense, and remediate POPs. One such example includes engineered antibodies that contain binding domains that can specifically associate with PCBs and other POPs, which can be utilized for specific binding and sensing applications. Due to inherent cost barriers and stability issues of these antibodies, though, other low-cost, stable engineered materials have been sought as an alternative platform to mimic this specific binding interaction; growing data are implicating the use of nutrient polyphenols as effective components of engineered POP capture/sensing platforms both due to their fluorescent detection characteristics and affinity for chlorinated organics. It has been reported that the binding domains of PCB-specific antibodies (e.g., the monoclonal antibody S2B1) form sterically constrained, highly aromatic pockets, which allow for pi-pi bond stacking interactions. Researchers have successfully mimicked these interactions in the development of synthetic materials, including those composed of polyphenolic moieties, for capture and sensing applications. For instance, by exploiting their biomimetic binding properties, researchers have developed porphyrin-containing compounds (highly aromatic compounds with similar molecular geometries to polymeric polyphenols) for the detection of environmental contaminants.

There is emerging evidence that biomimetic PCB binding domains can be synthesized by incorporating phenolic and related moieties into polymeric coatings, which will greatly enhance the ability to detect and remove/remediate PCB contamination Polyphenols such as quercetin and curcumin are of special interest not only because of their beneficial health effects and environmentally benign nature but also because of their highly aromatic structures, which may enable effective and/or specific binding and sensing of aromatic pollutants. Recently, we developed a novel strategy to synthesize polymeric networks that incorporate these polyphenols as both stable coating materials and biodegradable hydrogels for use as a pollutant binding platform and an antioxidant release platform to combat localized cellular oxidative stress, respectively. These materials can have enhanced biocompatibility over their free compound form and in preliminary studies, PCB binding is increased through the incorporation of phenolic moieties in a polymeric matrix. Additionally, the characteristic fluorescence peaks associated with these flavonoids effectively have been used preliminarily to sense PCB congeners in a concentration-dependent fashion.

Although, biomimetic polymeric networks have been utilized for rapid binding of relevant pollutants, these pollutant-bound materials must be easily removed from contaminated environments in order to be effective remediation systems. Combining polyphenolic polymers with magnetic nanoparticles in nanocomposite microparticle systems, then, allows for rapid binding of aromatic pollutants followed by high-throughput magnetic separation from contaminated samples. Magnetic nanoparticles (MNPs) have found widespread applications in magnetic fluids, catalysis, and bioseparation, and are being utilized increasingly in environmental remediation, either as uncoated particles or polymer-functionalized particles for increased pollutant adsorption, sensing and degradation. Although nanoparticles in general have been studied for some time to elucidate their potential toxicity, magnetic iron oxide nanoparticles have shown minimal cytotoxicity and additionally have been used successfully in biomedical applications for many years in areas such as magnetic resonance imaging, targeted drug delivery, and hyperthermia. Polyphenolic polymers can enhance the biocompatibility of metal nanoparticles like iron oxide, thus forming the basis for a safe, environmentally-sound remediation platform. MNPs have found great utility in biomedical applications due to their superparamagnetic properties (conferred by their nanoscale size), which allow for easy manipulation with a magnetic field (e.g., for effective magnetic separation of captured molecules or for transport and immobilization at target sites) and heat generation in response to an alternating magnetic field (AMF) for local heating to modify properties (e.g. reaction rates, binding properties, thermal treatment). By utilizing diet-derived polymeric polyphenol coatings of defined thickness and matrix structure, Pollutants can be very quickly removed from contaminated water samples. Further, heating of the material can result in thermal destabilization of the polymer matrix and subsequent release of bound pollutants for potential reuse. This combined system, incorporating the binding potential of nutrient polyphenols with the separation and remote heating potential of magnetic nanoparticles potentially could be used for rapid removal of pollutants from environmental samples. The addition and implementation of bioactive nutrients into engineered polymers and nanoparticles allows for efficient remediation of contaminated sites and is worthy of increased investigation due to the high efficiency and specificity potential as well as the environmentally responsible makeup of the technology.

REMEDIATION OF POP USING GENETICALLY MODIFIED PLANTS

The term Persistent Organic Pollutants (POPs) is used to describe a group of anthropogenic, often lipophilic, chemicals that are toxic and recalcitrant to degradation in the environment. Some of the more volatile POPs, such as polychlorinated biphenyls (PCB) and poly-brominanted diphenyl ethers (PBDEs), can freely disperse around the globe, to areas that are geographically distant from the source. Furthermore, many of these POPs bioaccumulate, the PBDE 2,3-dibromopropyl-2,4,6-tribromophenyl ether (DPTE) has been detected in harp seals (Pagophilus groenlandicus) from the Barents and Greenland Seas at concentrations up to 470 nanograms per gram wet weight in blubber.

Many of the POPs were used extensively between the 1930s and 1980s worldwide, when knowledge of the long-term toxicity was unknown. In response to growing global concern, in 2001 a global treaty, The Stockholm Convention on Persistent Organic Pollutants led by the United National Environmental Programme, was ratified by the international community, with the aim of eliminating or restricting the production and use of 12 of the most damaging POPs. The listed POPs comprised pesticides, industrial chemicals and unintended industrial by-products. Nearly all of the compounds listed in these two tables are classified as possible or probable human carcinogens (respectively, the International Agency for Research on Cancer, IARC, and the United States Environmental Protection Agency, US EPA), and exert additional toxicities such as endocrine disruption. Prominent industrial chemicals on the list include PCBs and hexachlorobenzene. Mixtures of these chlorinated hydrocarbons were used in the electrical industry, as well as in heat transfer fluids, plasticizers, lubricants and paints. The pesticides listed were, in the early 20th century, considered beneficial and widely used with naivety as to the longer-term health risks; for example, dichlorodiphenyl trichloroethane (DDT) was extensively used in the fight against malaria.

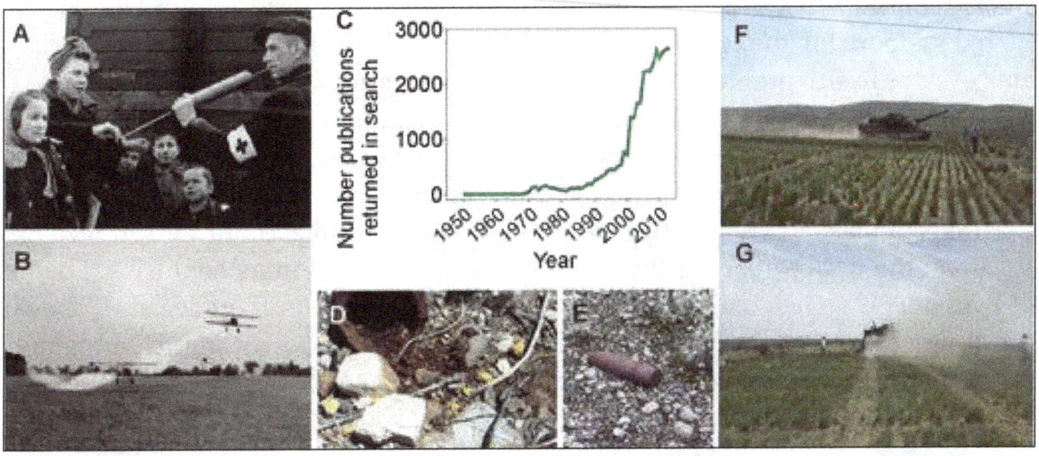

Use of persistent organic pollutants: (A) Three dusting planes spraying DDT over an open field. (B) Delousing with DDT, Poland, 1946. (C) Number of publications in National Center for Biotechnology Information search for 'phytoremediation'. (D) Explosives residues from a 155-mm Howitzer round. (E) Unexploded ordnance on a military training range. (F) and (G), testing resistance of grass species to military vehicles, Yakima Training Center, Washington, USA.

Table: Persistent organic pollutants agreed by member states following The Stockholm Convention treaty 2001, to be eliminated, or phased out.

Name	Chemical Abstracts Service (CAS) Number	Class	Target
Aldrin	309-00-2	Pesticide	Elimination
Chlordane	57-74-9	Pesticide	Elimination
Dieldrin	60-57-1	Pesticide	Elimination
Endrin	72-20-8	Pesticide	Elimination
Heptachlor	76-44-8	Pesticide	Elimination
Hexachlorobenzene (HCB)	118-74-1	Industrial chemical	Elimination
Mirex	2385-85-5	Pesticide	Elimination
Toxaphene	8001-35-2	Pesticide	Elimination
Polychlorinated biphenyls (PCBs)	Various	Industrial chemical	Elimination
Dichlorodiphenyl trichloroethane (DDT)	50-29-3	Pesticide	Restriction
Polychlorinated dibenzo-p-dioxins ('dioxins')	Various	Unintentional by-product	Unintentional Production
Polychlorinated dibenzofurans	Various	Unintentional by-product	Unintentional Production

Production and use of nearly all of the pesticides listed has now been phased out in member states. The use of some POPs continues however, because in the absence of suitable replacements, their use outweighs the risks, the World Health Organization declared support for the indoor use of DDT in African countries, where malaria remains a major health problem. Moreover, some POPs are released unavoidably into the environment, for example, significant dibenzodoxines and dibenzofurans are produced during the incineration of coal, wood, peat and waste from domestic fires, in vehicle emissions and natural events, such as volcanic eruptions and forest fires. Also, despite legislations in place to prevent release, accidental spills still account for significant releases, and the use of many of the POPs continues in developing countries. The long half-lives of POPs mean that significant pollution can still be present decades following the last use. For example, levels of POPs, including DDT, polybrominated biphenyls and hexabromobenzene, released from the Michigan Chemical plant in St. Louis, Michigan, are still contaminating and adversely affecting wildlife 40 years later. In addition, emerging pollutants such as pharmaceuticals in soil and water, are now also of concern. There are a range of remediation methods, reviewed by Marmiroli *et al*, which include incineration, chemical treatment and aerobic and anaerobic biodegradation in composting. However, many current treatments require the removal of contaminated land for treatment, and on larger sites treatment costs can become prohibitive. Phytoremediation, as an *in situ* method, is particularly suited to large-scale sites. Previous comprehensive reviews in this area have covered earlier progress. More recently, along with a continuation in the increasing number of scientific publications, focus is now turning from model plant systems to the transfer of this technology into plant species suitable for remediation in the field. Moving into non-model systems raises new challenges: the development of transformation techniques and identification of reliable transgene promoters. Questions are also raised concerning public acceptance and financial feasibility, with GM-based phytotechnologies facing costly and time-consuming legislations.

Table: Persistent organic pollutants subsequent added to the 2012 Stockholm Convention treaty by Conference of Parties up to 2013.

Name	CAS Number	Use	Target
Chlordecone	143-50-0	Pesticide	Elimination
Endosulfan	115-29-7959-98-833213-65-9	Pesticide	Elimination
Hexabromo-biphenyl	36355-01-8	Industrial chemical	Elimination
Hexabromo-diphenyl ether and Heptabromo-diphenyl ether	Various	Industrial chemical	Elimination
Hexabromo-cyclododecane	25637-99-43194-55-6134237-50-6134237-51-7 134237-52-8	Industrial chemical	Elimination
β-Hexachloro-cyclohexane	319-85-7	Pesticide	Elimination
Lindane (gamma-hexa-chloro-cyclohexane)	58-89-9	Pesticide	Elimination
Pentachlorobenzene	608-93-5	Industrial chemical and by-product	Elimination and unintentional production
Perfluorooctane-sulfonic acid (PFOS), its salts and perfluorooctane-sulfonylfluoride (PFOSF)	Various	Industrial chemical	Restriction
The poly-brominanted diphenyl ethers (PBDEs): tetrabromo-diphenyl ether and pentabromo-diphenyl ether	Various	Industrial chemical	Elimination

Microbial Degradation for in Situ Remediation of Pops

Microorganisms have a phenomenal capacity to evolve catabolic pathways for deriving energy from surrounding organics, and the genetic diversity of soil microbiomes has accounted for the discovery of catabolic pathways for hundreds of organic compounds. On the face of it, microbial biodegradation presents a natural, holistic approach and microbes with the ability to degrade specific POPs can often be isolated from soil contaminated with that particular pollutant, then enriched cultures reintroduced. However, for successful bioremediation, the environment has to be favourable for the establishment of POP-metabolizing microbes; this can be hindered by a range of environmental factors such as suboptimal soil pH or temperature and the presence of co-pollutants, for example, other POPs or toxic metals.

Phytoremediation for In Situ Remediation of Pops

Phytoremediation is described by the US EPA as 'the use of plants to partially or substantially remediate selected contaminants in contaminated soil, sludge, sediment, ground water, surface water, and waste water'. There is a rapidly increasing number of scientific publications relating to this term and an expanding number of ways in which plants can be used for effective exposure prevention. The presence of plants in terrestrial systems can provide phytostabilization, reducing loss of valuable top soil, which is ecologically and aesthetically desirable. Furthermore, the presence of a diverse plant community can promote microbial numbers and diversity, and plants can induce POP-detoxifying activity in microbiota and contribute to the co-metabolism of xenobiotics Different soil types can affect the bioavailability of the pollutant, and for the pollutant to be taken up by the plant it needs to be moderately hydrophobic, defined as having an octanol:water partition coefficient ($\log K_{ow}$) of 0.5–3.0. However there are a number of limitations to using plants for remediation: it takes time for plants to grow and establish, and also there are costs associated with planting, cultivation and contaminant monitoring.

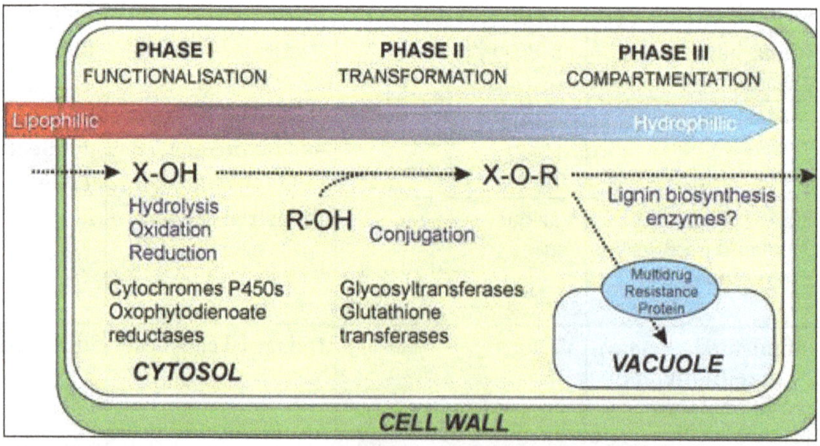

Classical pathway for xenobiotic detoxification in plants. In Phase I the foreign compound is functionalized, enabling Phase II, conjugation to sugars or glutathione, which facilitate Phase III, compartmentation to the vacuole or apoplast.

The classical pathway for xenobiotic detoxification in plants is outlined in figure. Following uptake, one or more functional groups on the molecule are transformed, often by cytochromes P450. Transformation enables the second step, conjugation to occur, usually resulting in a more polar, less toxic compound. This step is often catalysed by glutathione transferases or glycosyltransferases. The final step is transportation of the xenobiotic conjugate to a less biochemically sensitive location such as the vacuole, incorporation into macromolecular structures such as lignin, or sequestration back into the environment, often as volatilization. Although plants possess a wealth of enzymes for xenobiotic detoxification, a major constraint for phytoremediation is the low levels of specific *in planta* catabolic activity to fully degrade target POPs. The application of GM can go some way in mitigating this limitation. By transferring microbial genes relaying xenobiotic-catabolism into selected plant species, detoxification, or, ideally, mineralization of organic pollutants can be achieved.

Genetic Modification Approach

Transgenic approaches to PCB phytoremediation are more challenging than in the other case, due to the diversity of co-polluting PCBs and the fact that multiple genes need to be introduced for complete catabolism. Initial attempts to stably transform BphA1/BphA2, BphA3 and BphA4 (which catalyse aerobic oxygenation of biphenyl) into tobacco, failed to simultaneously express all components. Successful expression of a single component can have a substantial effect on PCB remediation however, as demonstrated by the enhanced resistance to PCBs of tobacco transformed with bphc; this gene encodes 2,3-dihydroxybiphenyl-1,2-dioxygenase, which catalyses ring cleavage in the biphenyl degradation pathway. The prior understanding of microbial PCB catabolism, coupled with recent advances in multiple transgene expression systems, demonstrate that the assembly of degradative pathways in plants is a now realistic target.

GM Technologies

Public Perception

While GM technologies could be a powerful tool in the remediation of POPs, the promising results outlined above are crucially dependent upon public acceptance. In 2012, 180 million hectares of arable land worldwide were used to plant GM crops, suggesting widespread public acceptance of GM technologies. However, there is still significant and global public anxiety regarding the release of GM organisms, with particular concern within Europe. Data show that GM crops contribute substantial economic and environmental benefits, but the absence of any direct benefit for the consumer likely contributes to the poor public acceptance of these crops. Next generation GM crops targeting major chronic diseases (obesity, heart disease, type 2 diabetes) are being licenced for commercialization in the next five to ten years and these health-giving properties that might contribute to future public acceptance.

In the case of GM plants designed to remediate toxic organic pollutants, there is understandable concern surrounding the potential accumulation of these compounds and metabolites in plants and, through herbivory, the detrimental effects that this could have on food chains. The best scenario for a GM-based remediation approach is the complete mineralization of the POP, with no accumulation of toxic intermediates. Such a scenario is potentially the case for the remediation of RDX outlined above. Research indicates that public understanding of science is key to engaging support for GM technologies. The remediation of RDX represents a phytoremediation standard that, if effective in the field, could be used to promote greater public acceptance of GM crops used on contaminated soil.

Costs

A major shortcoming in the advancement of GM-based technologies for phytoremediation is the cost and time taken to test, licence and field-trial GM plants. The costs of complying with GM regulations for the release of an insect-resistant GM maize in the USA were calculated to be between €5.5M and €11.5M. In addition, the time taken to push an application through the review process can significantly impact on the costs and marketing. In comparison with these legislative costs, a report commissioned by the UK Department for Environment, Food & Rural Affairs calculated remediation costs based on estimates given in a questionnaire by UK contaminated land organizations. For areas greater than 5000 m^3, the combined costs of excavation and landfill disposal of a hectare to a depth of one metre were calculated as between £32 000 and £520 000. Costs for phytoremediation, although based on a single respondent, were estimated at £5 000 to £10 000; between approximately six and 50 times less than for excavation and landfill, and within the 20 times less value reported by Paz-Alberto et al, suggesting that, assuming the costs of licencing can be borne, the use of GM-based remediation could be a cost-effective approach. It is clear that the cost and timescales required for regulatory approval mean that GM-based remediation technologies are currently prohibitive for all but large companies and state-funded projects. As a result of this it is likely that projects would initially be restricted to the remediation of more commonly-occurring xenobiotics, or ones for which clean-up is in higher demand, and where the same GM plant line can be used at multiple sites.

Areas for Advancement

Identification and Optimizing Genetic Resources for POP Detoxification

There are many chemically different POPs, but GM-technologies enable phytoremediation to be tailored towards a specific pollutant. The number of microorganisms isolated with abilities to degrade POPs is gathering momentum, including those for PCBs (heptachlor), DDT and organochlorinated insecticides [such as aldrin, chlordecone, dieldrin, endosulfan and lindane. Metagenomics studies have identified genes encoding much of the DDT degradation pathway and a study has shown that plants can take up DDT. The next step is to combine DDT-degrading genes in model plant species, such as

Arabidopsis, to test the activity of this pathway in plants. Similar technologies are now also enabling the complex relationships between plants and associated microbial communities to be studied, with the potential for discovering new inter-species pathways for POP catabolism. A critical factor for the successful remediation of POPs is the complete breakdown of the compounds. For many POPs, such as TNT and PCBs, further biochemical steps need to be identified or developed.

Molecular Biology Techniques

The GM approaches mostly concentrate on the use of one or two transgenes to confer a xenobiotic-detoxifying activity. This approach relies on harnessing endogenous plant enzymes for subsequent downstream catabolism and additionally, in field conditions, a range of POP pollutants with toxic effects may be present. While transferring multigene pathways has been achieved e.g. for dichloroethane, it is challenging, relying on co-transformation with distinct vectors containing separate marker genes. Trait stacking can now be more readily achieved using vectors designed for multi-gene transformation, such as GoldenGate and GoldenBraid. Using classical transformation techniques, it is not possible to predict where the transgene will insert into the plant genome, with critics arguing that there could be unintended consequences resulting from this. New methods using genome editing technologies such as ZFN, TALEN and CRISPR are enabling transgenes to be inserted into specific locations in the genome at sites that do not disrupt the endogenous activities of the plant.

Future work should also concentrate on sequencing the genomes of remediation relevant species such as miscanthus, vetiver and giant reed. The production of annotated genetic maps for these species will enable the use of genome editing technologies. Characterized dicot, viral and even other monocot promoters may not work as well in these monocot species, and sequencing data can be used to identify promoters contributing cis-elements and codon bias from the host plant to ensure high transgene expression. Transgene expression could also be targeted to the in planta location of the xenobiotic by the use of organ-specific promoters, reducing the metabolic load of transgene expression.

Field-applicable Species

Many of the bench-scale GM studies have been performed in Arabidopsis or tobacco, and this technology is now being transferred into more phytoremediation applicable species. Each contaminated site will require specific features from the remediating plant species. For example, military training ranges require plants that do not obstruct training operations, and thus favour grasses over trees. Species are also required to be fire-resistant, and to be able to withstand the movement of heavy military vehicles. Classic phytoremediation species have attributes such as fast growth rates, high biomass and higher intrinsic tolerance to POP and co-contaminant toxicity. Candidate species include members of the willow (Salicaceae) family such as willow, poplar, aspen and cottonwood, and a screen of willow and poplar has now identified varieties with high in situ phytoremediation potential. Where trees are not suitable, grasses such as the non-invasive hybrid miscanthus Miscanthus × giganteus, switchgrass, vetiver grass (Vetiveria zizanioides) or the wetland species giant reed (Arundo donax) possess desirable traits. Additional species, including ones suitable for aquatic and saline environments. A further advantage of some of these species is that they can feed into the biorefinery and bioenergy industries, with agricultural practices already in place for cultivation and harvesting. Significant advances are already being made in optimizing the transformation technology for monocot species such as maize, and studies in bioenergy-related areas have resulted in the development of transformation protocols for species such as switchgrass but future research efforts are necessary for species such as miscanthus, vetiver and giant reed, where gene transfer and plant regeneration techniques are not yet effective.

PERMISSIONS

All chapters in this book are published with permission under the Creative Commons Attribution Share Alike License or equivalent. Every chapter published in this book has been scrutinized by our experts. Their significance has been extensively debated. The topics covered herein carry significant information for a comprehensive understanding. They may even be implemented as practical applications or may be referred to as a beginning point for further studies.

We would like to thank the editorial team for lending their expertise to make the book truly unique. They have played a crucial role in the development of this book. Without their invaluable contributions this book wouldn't have been possible. They have made vital efforts to compile up to date information on the varied aspects of this subject to make this book a valuable addition to the collection of many professionals and students.

This book was conceptualized with the vision of imparting up-to-date and integrated information in this field. To ensure the same, a matchless editorial board was set up. Every individual on the board went through rigorous rounds of assessment to prove their worth. After which they invested a large part of their time researching and compiling the most relevant data for our readers.

The editorial board has been involved in producing this book since its inception. They have spent rigorous hours researching and exploring the diverse topics which have resulted in the successful publishing of this book. They have passed on their knowledge of decades through this book. To expedite this challenging task, the publisher supported the team at every step. A small team of assistant editors was also appointed to further simplify the editing procedure and attain best results for the readers.

Apart from the editorial board, the designing team has also invested a significant amount of their time in understanding the subject and creating the most relevant covers. They scrutinized every image to scout for the most suitable representation of the subject and create an appropriate cover for the book.

The publishing team has been an ardent support to the editorial, designing and production team. Their endless efforts to recruit the best for this project, has resulted in the accomplishment of this book. They are a veteran in the field of academics and their pool of knowledge is as vast as their experience in printing. Their expertise and guidance has proved useful at every step. Their uncompromising quality standards have made this book an exceptional effort. Their encouragement from time to time has been an inspiration for everyone.

The publisher and the editorial board hope that this book will prove to be a valuable piece of knowledge for students, practitioners and scholars across the globe.

INDEX

CPSIA information can be obtained
at www.ICGtesting.com
Printed in the USA
BVHW062118290822
645775BV00003B/185

9 781639 874187